Helmut = IKAN

Rudie =

Luzón
Manila

PACIFIC OCEAN

PHILIPPINES
Cebu

Palawan

SULU SEA

Mindanao

Palau Is.

Sipadan Is.

CELEBES SEA

Talaud Is.

Derawan Is.   Manado   Halmahera

Biak

Sulawesi

N   E   S   I   A

Irian Jaya

Ambon

BANDA SEA

New Guinea

Flores

Timor

© 1994 Tetra-Press
Tetra-Werke Dr. rer. nat. Ulrich Baensch GmbH
Herrenteich 78, D-49324 Melle, Germany

All rights reserved, incl. film, broadcasting,
television as well as the reprinting

Distributed in U.S.A. by
Tetra Sales U.S.A.
3001 Commerce Street
Blacksburg, VA 24060

Distributed in UK by
Tetra Sales, Lambert Court,
Chestnut Avenue, Eastleigh Hampshire S05 3 ZQ

WL-Code 16586
ISBN 1-564-65-170-3

Printed in Germany

Rudie H. Kuiter / Helmut Debelius

SOUTHEAST ASIA
TROPICAL FISH GUIDE

**Indonesia   Philippines   Vietnam   Malaysia
Singapore   Thailand   Andaman Sea**

Over 1000 photographs
of marine fishes
taken in their natural habitat

TABLE OF CONTENTS

CLASS CARTILAGINOUS FISHES	CHONDRICHTHYES	9
Family Six & Sevengill Sharks	Hexanchidae	10
Family Whale Sharks	Rhincodontidae	10
Family Bullhead Sharks	Heterodontidae	11
Family Wobbegong Sharks	Orectolobidae	11
Family Bamboo Sharks & Epaulette Sharks	Hemiscylliidae	12
Family Zebra Sharks	Stegostomatidae	14
Family Nurse Sharks	Ginglymostomidae	14
Family Sand Tiger Sharks	Odontaspididae	15
Family Mackerel Sharks	Lamnidae	15
Family Thresher Sharks	Alopiidae	16
Family Cat Sharks	Scyliorhinidae	16
Family Requiem Sharks	Carcharhinidae	17
Family Hammerhead Sharks	Sphyrnidae	22
Family Angel Sharks	Squatinidae	23
Family Sawfishes	Pristidae	24
Family Sharkfin Guitarfishes	Rhynchobatidae	25
Family Torpedo Rays	Torpedinidae	26
Family Stingrays	Dasyatididae	26
Family Stingarees	Urolophidae	31
Family Cownose Rays	Rhinopteridae	31
Family Eagle Rays	Myliobatididae	32
Family Devil Rays	Mobulidae	33

CLASS BONY FISHES	OSTEICHTHYES	34
ORDER	ANGUILLIFORMES	
Family Morays	Muraenidae	34
Family Snake Eels	Ophichthidae	40
Family Garden Eels	Heterocongridae	41
ORDER	CLUPEIFORMES	
Family Herrings	Clupeidae	42
Family Anchovies	Engraulidae	42
ORDER	SILURIFORMES	
Family Eeltail Catfishes	Plotosidae	43
ORDER	AULOPIFORMES	
Family Lizardfishes	Synodontidae	44
Family Grinners	Harpadontidae	46
ORDER	MYTOPHIFORMES	
Family Lanternfishes	Myctophidae	47
ORDER	OPHIDIIFORMES	
Family Lings	Ophidiidae	48
ORDER	LOPHIIFORMES	
Family Anglerfishes	Antennaridae	49
ORDER	GOBIESOCIFORMES	
Family Clingfishes	Gobiesocidae	52
ORDER	BELONIFORMES	
Family Long Toms	Belonidae	54

Family Garfishes	Hemiramphidae	54
Family Flyingfishes	Exocoetidae	55
ORDER	BERYCIFORMES	
Family Flashlightfishes	Anomalopidae	56
Family Squirrel & Soldierfishes	Holocentridae	57
Subfamily Squirrelfishes	Holocentrinae	57
Subfamily Soldierfishes	Myripristinae	62
ORDER	GASTEROSTEIFORMES	
Family Seamoths	Pegasidae	64
ORDER	SYNGNATHIFORMES	
Family Trumpetfishes	Aulostomidae	65
Family Flutemouths	Fistulariidae	66
Family Shrimpfishes	Centriscidae	67
Family Ghost Pipefishes	Solenostomidae	68
Family Pipefishes & Seahorses	Syngnathidae	71
Subfamily Seahorses	Hippocampinae	71
Subfamily Pipefishes	Syngnathinae	73
ORDER	SCORPAENIFORMES	
Family Scorpionfishes	Scorpaenidae	76
Family Waspfishes	Tetrarogidae	84
Family Flying Gurnards	Dactylopteridae	85
Family Flatheads	Platycephalidae	86
ORDER	PERCIFORMES	
Family Rockcods, Groupers & Basslets	Serranidae	88
Subfamily Groupers & Cods	Serraninae	88
Subfamily Basslets & Seaperches	Anthiinae	99
Subfamily Soapfishes	Grammistinae	108
Family Dottybacks	Pseudochromidae	112
Family Longfins	Plesiopidae	116
Family Bigeyes	Priacanthidae	118
Family Cardinalfishes	Apogonidae	120
Family Tilefishes	Malacanthidae	127
Family Cobias	Rachycentridae	130
Family Remoras	Echeneidae	130
Family Jacks & Trevallies	Carangidae	132
Family Snappers	Lutjanidae	137
Family Fusiliers	Caesionidae	146
Family Silverbellies	Gerreidae	149
Family Sweetlips	Haemulidae	150
Family Emperors	Lethrinidae	155
Family Spinecheeks	Nemipteridae	158
Family Goatfishes	Mullidae	161
Family Bullseyes	Pempherididae	166
Family Drummers	Kyphosidae	168
Family Batfishes	Ephippidae	170
Family Silver Batfishes	Monodactylidae	173
Family Scats	Scatophagidae	173
Family Butterflyfishes	Chaetodontidae	174
Family Angelfishes	Pomacanthidae	187
Family Damsel & Anemone Fishes	Pomacentridae	198

Family Hawkfishes	Cirrhitidae	212
Family Bandfishes	Cepolidae	215
Family Jawfishes	Opistognathidae	215
Family Mullets	Mugilidae	217
Family Wrasses	Labridae	218
Family Parrotfishes	Scaridae	235
Family Sand Divers	Trichonotidae	239
Family Grubfishes	Pinguipedidae	240
Family Convict Blennies	Pholidichthyidae	242
Family Triplefins	Trypterygiidae	242
Family Blennies	Blenniidae	244
Family Dragonets	Callionymidae	250
Family Gobies	Gobiidae	253
Family Dart Gobies	Microdesmidae	265
Family Rabbitfishes	Siganidae	269
Family Moorish Idols	Zanclidae	273
Family Surgeon & Unicornfishes	Acanthuridae	274
Family Barracudas	Sphyraenidae	286
Family Tunas & Mackerels	Scombridae	288
Family Billfishes	Istiophoridae	291
ORDER	PLEURONECTIFORMES	
Family Lefteyed Flounders	Bothidae	292
Family Soles	Soleidae	293
ORDER	TETRAODONTIFORMES	
Family Triggerfishes	Balistidae	294
Family Filefishes	Monacanthidae	300
Family Boxfishes	Ostraciidae	305
Family Pufferfishes	Tetraodontidae	308
Family Porcupinefishes	Diodontidae	315

Philippine fishermen chasing in formation.

INTRODUCTION

The region of Southeast Asia includes over 20,000 islands, countless coral reefs, and in addition there is a mixture of faunas from different oceans. The equator crosses along the middle and currents run complicated courses which changes with tides and seasons. These are some of the ingredients supporting the richest marine fauna of the world.

This book serves as a companion volume to the INDIAN OCEAN TROPICAL FISH GUIDE, Debelius, 1993, which covered the vast expand of the Western Indian Ocean. Our geographical area includes the Eastern Indian Ocean, Andaman Sea, Thailand, Malaysia and Indonesia to the Philippines, including the South China Sea area to Vietnam. This large geographical area has an enormous variety in habitats which can be appreciated by just looking at the map. Along the south is the Indian Ocean, were a large swell is rarely

Exciting diving at the Talaud-Islands, northern Indonesia.

absent, with high energy zones, and to the north to the calmest waters all year round between large land-masses which range from large shallow flats to some very deep trenches. As fishes are very habitat specific, there is a great variety in our area. Just reef-fishes alone we estimate more than 3000 species.

Underwater faunas can differ greatly between different areas, including from one island or indeed from one reef to the next. Just like on land where water or mountains divide the fauna, underwater the currents, depths and landmasses separate faunas as effectively. Over time the water levels changed greatly, especially during ice-ages, dropping more then 100 m during certain periods which meant that you could walk to Bali from China, and in addition temperatures also altered. This effected the fauna in several ways. During the time the sea level was very low, a once single fauna became divided between Indian and Pacific Ocean which in time caused changes in many of the fish species. Some changed more than others and we call them geographical variations, sub-species or sibling species. Many of these co-occur now the sea-levels are high again and the faunas can mix to certain extend. Temperature changes effects the fauna as a level suits particular species. When temperatures change the fauna migrates to the adjacent areas becoming more suitable, which can be in several directions. Thus can split the fauna with species no longer existing in the original zone and comprising isolated populations. If migration is not possible than adapting to changing demands are necessary to survive. If changes are gradual, many species can adapt but if changes are quick there will be a drop-out of many species. With large land masses migration is usually possible as temperatures change, the fauna can remain in touch with the right conditions, but a small isolated island offers no alternatives than adapting. This is one of the main reasons for low diversity on remote outcrops. It has also caused colour variations in wideranging species, especially between populations along the continental margins and those more oceanic.

Obviously temperatures play an important role to limit the range of particular species and this is often demonstrated in our area with the fishes. In general in our area the Indian Ocean is lower in temperature by several degrees compared Pacific, and the range limit of the fauna is very distinct in some places. The Indonesian Archipelago is distinctly Indian Ocean along the southern coasts as far as Timor. On the north coasts of the southern chain of islands there is a mix of Indian and Pacific faunas between Java and Bali. We estimate only about 10% of Java's north coast is of Pacific origin, but in contrast about 10% of Bali's north coast is Indian fauna. This may vary over time as conditions change. On a recent trip to the south coast of Java special attention was given to Pacific and Indian Ocean species or forms, and it was a surprise that not one Pacific fish could be found as it disagrees with some surveys. No doubt this is due to the poor knowledge of the fauna in general, and the reporting of Pacific species for the Indian siblings. Sumatra is one area where few fish-experts have been, and no doubt it will have many surprises in store, as Java did with many new records for Indonesia, including some species until then only known from Andaman Sea and even Maldive Islands.

At Sipadan, the famous dive-island in the Celebes Sea.

In the north, the marine faunas are continuations along the continental margins, island chains and ridges which embraces a number of seas. The South China Sea ranges from north of the Philippines along Vietnam and extending south to Malaysia and Singapore. The fauna along its continental margin ranges from southern Japan to Java. The Philippines fauna is more oceanic with similar species ranging along the east from southern Japan to New Guinea. Between the South China Sea and Pacific Ocean there are some large water masses: the Sulu Sea, Celebes Sea and one of the most diverse areas the Banda Sea which has one of the deepest areas in our area, down to 7440 m. Each sea though has its own unique species make up and each is a species bonanza.

Bali, 'the' relation between culture, tradition and diving.

Until recently when scuba diving became popular in our area, little scientific work was done since a Dutch army doctor Bleeker almost 150 years ago. Interestingly the first scientific collection reported by P. Bleeker was of Flores where we did our first and most diving, and the collecter named J. C. J. Hellmuth. Bleeker published 500 papers with an incredible 406 new genera and 3,324 species, including freshwater of which 75% devoted to the Indonesian fauna. This became the base of his famous ATLAS ICHTHYOLOGIQUE with most perfect drawings of the fishes in detail. Many of the species he described were not seen until recent times and some were found by going to the original type localities.

The most popular dive-area in the Andaman Sea are the Similan-Islands.

As we have the advantage of looking at fish at first hand in their environment, we can see the interreaction of species between different sexes or similar species. Some species have completely different colour forms between sexes and also between small juveniles and adults. Bleeker sometimes suspected the possibility of some of the species being of different sex but it was difficult to prove by having specimens mostly collected by fisherman or from markets.

Now with about 20 years of active diving in south east Asia and a rapidly growing sport, the situation with fishes has come to the point that we have a fair idea what is there, however there are many new species, similar species with some confusion, and no doubt more will be

Muro-Ami-children diving in the Sulu Sea, Philippines.

found. Some species were recently described and a large number are in the process of being done. Work in ichthyology and taxonomy is never finishes, and more importantly we need to know about behaviour, ecology, habitats etc, thus about fishes.

This book is only a small window into the sea, viewing fishes we are likely to see, but not al could be included. Some more interesting replace those not so interesting or simply the more beautiful are preferred to the ugly. A choice had to be made as it is impossible to include every species in the size of a book such as this. It combines the common and obvious species with those as a family more important or just because they are special and spectacular. The main purpose of this book is to learn about fishes and encourage those with a camera to keep at the greatest challenge...photographing fishes.

Frankfurt, Autumn 1994   **Rudie H. Kuiter + Helmut Debelius**

METHODS AND EXPLANATION OF TERMS

CLASSIFICATION
The species are arranged in scientific order according to most recent authors. Within a family the most similar species follow each other and where possible are featured on the same page to make comparison easy. Each species has a scientific name which is in two parts, the genus followed by the specific name. The genus is the same between species which feature a certain features or characters, which is a level of difference. The closer the species is related, the least the difference. The level of difference increases as species are less related and certain levels are defined and termed. At the top is a Class, where all the sharks and rays are grouped in one and all the bony fishes in the other. Each of these has more defined groups, each referred to as an Order. Following the same principle an Order comprises Families, Families comprises Genera and the latter the Species. Not always are such levels clear or between levels and these are Sub-Order, Family etc.

The name of each species originates when it is scientifically described and published in special journals. It is assigned to existing family, genera and given a species name. Sometimes the genus or even family is new as well. When a species is published in a book such as here, the scientific name is followed by the name of the person who made the description and the date when this was published. Sometimes the genus name is changed over time and if this has occurred the person's name and date is placed in brackets to indicate this.

FISHES
There are over 22.000 fishes worldwide, about 13.000 of these marine. The two major groups are the Cartilaginous Fishes, the Sharks and Rays, and the Bony Fishes, comprising a great variety from eels to pufferfishes.

SHARKS AND RAYS
Shared features of the sharks and rays are the more or less hardened cartilagenous skeleton and for reproduction internal fertilisation. The sharks differ primarily from the rays in having pectoral fins separate from the sides of the head. In rays these are fused to the sides of the head over the gills and usually the groups are easily separated by general shape. There are at least 43 families, about 350 species of shark, and 500 rays.

BONY FISHES
All fishes which have a bony skeleton are included. Only the Coelacanths and Lungfishes are classified into a subclass, Sarcopterygii, as they have lobed fins, and all others in the subclass Actinopterygii, which have rayed fins. Only the latter is included here. They show a great range in diversity with sizes from about 1 cm to over 5 m fully grown, and shapes from spagetti like eels through seahorses to balloon like pufferfishs or shaped as a disc, leaf, dart, etc. The various distinct groups are in orders, families and genera, in all about 20.000 species.

TERMS USED
Length: the maximum size known, measured from tip of snout to and of tail, excluding soft filaments. Length varies between different areas, usually fishes grow larger in cooler conditions. For the rays this is not practicle and instead we use Width. **Distribution:** the total known range. Wideranging Indo-pacific includes the entire area covered by this book (our area). Some showing limited distribution undoubtly occur in adjacent waters, but not yet recorded from there. **Depth:** Based on our observations and liturature records. **General:** covers habitat, behaviour and aspects of interest.

SIX- & SEVENGILL SHARKS • HEXANCHIDAE

Bluntnose sixgilled shark

Length: To 4.8 m.
Distribution: Very patchy, globally, tropical to temperate. In our are known from Andaman Sea, Sumatra and northern Philippines.
Depth: 25-1000 m.
General: A deep water species which occasionally is encountered by divers at night in moderate depths. Variable, grey to brown. A single dorsal fin, placed well back on the body. 6 gill slits.
Feeds on other sharks, bony fish, squid and crustaceans.

Hexanchus griseus (Bonnaterre, 1788)

WHALE SHARKS • RHINCODONTIDAE

Whale shark

Length: 12+ m.
Distribution: Cosmopolitan in all tropical to warm-temperate seas.
Depth: Surface to 130+ m.
General: The worlds largest fish. A harmless planktivore, primarily oceanic but often seen near reefs in pursuit of food. A very distinc species in colour (spotted dorsal pattern) and shape with mouth more forward and broad compared to other sharks. They are a migratory species usually observed singly but even when occurring in small aggregations it is difficult to tell underwater as they are spread out. The photo with the hawksbill turtle was taken near a pinnacle in the Andaman Sea.
Whalesharks feed on a great variety of schooling small pelagics, including crustaceans, cephalopods and fish.
Whale sharks are ovo-viparous. Embryos hatch from the egg case within the female reproductive tract and are subsequently born.

Page 9:
A rare photo of whale sharks swimming close together.

Rhincodon typus (Smith, 1828)

BULLHEAD SHARKS • HETERODONTIDAE

Japanese bullhead shark
L: To 1.2 m. Di: Southern Japan to East China Sea. Known also from the Philippines.
De: 20-100+m. Ge: A temperate species ranging south along the mainland in cooler water zones. Similar **H. zebra** (Gray, 1831) with more distinct bands (photo below) is tropical (known from Vietnam and Indonesia) but rarely seen because of preference for deep water.

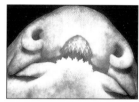

Heterodontus japonicus (Duméril, 1865)

WOBBEGONGS • ORECTOLOBIDAE

Tasselled wobbegong

Length: To 1.3 m.
Distribution: Eastern Indonesia, New Guinea and northern Australia.
Depth: 1-40+ m.
General: Head very broad and numerous branched tassels around mouth. During the day mainly inside narrow ledges, out at night. Like other wobbegongs they cannot be trusted and should not be handled as they can easily bite their own tail. In addition their mouth can lock on to anything that fits in the mouth. Mainly a coastal species, often in turbid waters and in still lagoons.

Eucrossorhinus dasypogon (Bleeker, 1867)

WOBBEGONGS • ORECTOLOBIDAE

Ornate wobbegong
L: To 3 m. Di: Eastern Indonesia, New Guinea and Australia. De: 1-40+ m. Ge: A large potentially dangerous species which should be left alone. Large specimens rest out in the open on top of reefs as well as in ledges. Clear coastal and offshore reefs, sometimes locally abundant.

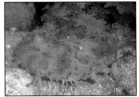

Orectolobus ornatus (de Vis, 1883)

BAMBOO & EPAULETTE SHARKS • HEMISCYLLIIDAE

Brown bamboo shark

Length: To 77 cm.
Distribution: From the Arabian Gulf into the Andaman Sea.
Depth: 5-80 m.
General: Inhabits lagoons and rocky shores.
Oviparous, lays up to 3 eggs cases on the bottom.
Feeds on crustaceans, fishes, and cephalopods.

Chiloscyllium griseum Müller & Henle, 1838

Grey bamboo shark
L: To 1 m. Di: Widespread tropical west Pacific. Containing all of our area. De: Intertidal to 85 m. Ge: Shallow reef flats to continental shelf. Secretive during the day, hiding in the back of ledges or under large and low table corals. Adults plain grey. Juveniles with broad dark bands.

Chiloscyllium punctatum Müller & Henle, 1838

BAMBOO & EPAULETTE SHARKS • HEMISCYLLIIDAE

White-spotted bamboo shark

Length: To 1 m.
Distribution: Continental from India to Japan, including all of our area.
Depth: 3-20 m.
General: A little known species, occasionally observed on reef crests at night. Sometimes in fish traps and often exported in the aquarium trade, apparently easily kept in captivity. Reported as living 25 years in one aquarium.

Chiloscyllium plagiosum (Bennett, 1830)

Epaulette shark
L: To 1.3 m. Di: Eastern Indian Ocean (photo from Cocos Keeling), New Guinea.
De: 1-10 m. Ge: A locally common species, including on intertidal coral reef flats, sometimes activily feeding around waders feet. A very distinct species with the large rounded black blotch just back and above the gills.

Hemiscyllium ocellatum (Bonnaterre, 1788)

Hooded epaulette shark
L: To 80 cm. Di: Indonesia and New Guinea. De: 3-20 m.
Ge: Secretive during the day. Crawles over reef crests during the night, hunting fishes and crustaceans. Distinct species, young with black snout and about 14 dark bands, integrated with small white spots. Banding fades in adults.

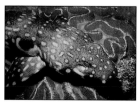

Hemiscyllium strahani Whitley, 1967

ZEBRA SHARKS • STEGASTOMATIDAE

Zebra shark
L: To 3.5 m. Di: Widespread tropical Indo-Pacific, including all of our area. De: 1-70 m. Ge: Small juveniles very attractive with strong dark bands, breaking up into spots with growth. Tail very long in adults. A nocturnal species, sometimes found resting on the substrate near reefs during the day.

Stegastoma fasciatum (Hermann, 1783)

NURSE SHARKS • GINGLYMOSTOMATIDAE

Tawny shark

Length: To 3.2 m.
Distribution: Widespread tropical Indo-Pacific, including all of our area.
Depth: 1-70 m.
General: Feeds primarily on octopus and other cephalopods. Nocturnal, resting in caves or large ledges, and often in groups, only occasionally moving about during the day. A large sluggish species, usually a yellowish colour with nasal barbels. May get aggressive if disturbed.
Juvenile spotted tawny shark left below

Nebrius ferrugineus (Lesson, 1830)

SAND TIGER SHARKS • ODONTASPIDIDAE

Grey nurse shark

Length: To 3.3 m.
Distribution: Globally in temperate to tropical zones, Vietnam, Philippines, in cooler zones of our region.
Depth: 10-200 m.
General: A locally common species, occurring in small aggregations along deep reefs, often in rocky gutters. One of the most fearsome looking sharks, but not aggressive towards divers. In some areas follows certain fish species such as sea-mullet along their migration path.

Carcharias taurus Rafinesque, 1810

MACKEREL SHARKS • LAMNIDAE

Shortfin mako

Length: To 1.3 m.
Distribution: Cosmopolitan, tropical and temperate seas, throughout our area.
Depth: 1-150 m.
General: One of the fastest and most travelled sharks, found mainly in waters with temperatures above 16 degrees Celsius. Like other members in the family their body temperature can be higher than that of surrounding waters, permitting a higher level of activity. A pelagic species, only occasionally encountered near reefs. Reports of the related great white shark or white pointer from our area are doubtful and therefore the species is not included here.

Isurus oxyrinchus (Rafinesque, 1810)

TRESHER SHARKS • ALOPIDAE

Tresher shark

Length: To 4 m.
Distribution: Circumtropical, throughout our area.
Depth: 1-400 m.
General: Coastal and off-shore waters. Body temperature higher than surrounding waters. A fast timid species, occasionally swimming along slopes or drop-offs adjacent to deep water. The very similar *A. pelagicus* is unlikely encountered in coastal waters and is mainly known from off-shore fishing with long-line. It differs in having shorter fins with more rounded tips, except tail which proportionally may even be longer.

Alopias vulpinus (Bonnaterre, 1788)

CAT SHARKS • SCYLIORHINIDAE

Coral cat shark

Length: To 70 cm.
Distribution: West Pacific and east Indian Ocean to India, ranging to New Guinea and Thailand, thoughout our area.
Depth: 1-15 m.
General: A strictly nocturnal species which is common but only seen at night on coral reef crests in coastal waters. Moves along the substrate, through narrow crevices in search for molluscs, shrimps and fishes. Easily recognised by the distinct colouration.

Atelomycterus marmoratus (Bennett, 1830)

Blotchy swell shark

Length: To 1.2 m.
Distribution: Mainly known from northern west Pacific, Japan to China Seas and possibly to New Guinea.
Depth: 20-200 m.
General: A nocturnal species, during the day on the bottom near reef. Mainly on rock substrate and usually in deep water, coming to shallower depths at night.
Feeds on a variety of benthic fishes.

Cephaloscyllium umbratile Jordan & Fowler, 1903

REQUIEM SHARKS • CARCHARHINIDAE

Silvertip shark

Length: To 3 m.
Distribution: Indo-Pacific, including all of our area.
Depth: 10-800 m.
General: A timid pelagic species, mainly in oceanic locations along very deep drop-offs, feeding on pelagic and demersal fishes. A fast and aggressive shark, reported to come in close to investigate noises. Young are more in-shore and tagging has shown that their range is fairly local-ised with most specimens re captured within 2 km of tagging site. This species is best recog-nised by the white posterior margins of the fins.

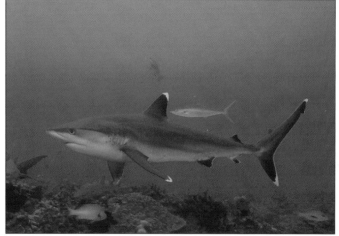

Carcharhinus albimarginatus (Rüppell, 1837)

Galapagos shark

Length: To 3 m.
Distribution: Cosmopolitan, tropical to temperate seas, but mainly associated with oceanic islands.
Depth: 1-180 m.
General: Distinct with large first dorsal fin and long ventrals with rounded tips. It prefers clear water and will come to the surface. A dangerous species and aggresive to divers, making threat displays as shown in the photograph. Taken in Andaman Sea where it is rare.

Carcharhinus galapagensis (Snodgrass & Heller, 1905)

17

REQUIEM SHARKS • CARCHARHINIDAE

Carcharhinus melanopterus (Quoy & Gaimard, 1824)

Blacktip reef shark

Length: To 1.8 m.
Distribution: Tropical Indo-Pacific, throughout our area.
Depth: 0.5-30 m.
General: Swims close to the substrate on reef crests, in channels and lagoons, entering very shallow depths and sometimes in brackish waters. Singly or in groups working together to hunt reef fishes over the shallow flats with dorsal fin out of the water, often targetting parrotfishes. A shy species towards divers in the shallows, seems less worried in deep sand lagoons.

Carcharhinus limbatus (Valenciennes, 1841)

Common blacktip shark

Length: To 2.6 m.
Distribution: Cosmopolitan, tropical to temperate seas, throughout our area.
Depth: 1-30 m.
General: Inhabits coastal waters, invading muddy estuaries and mangrove swamps. Distinct long snout and bronze upper surface. Black tips on most fins. 1-10 young which are born in extremely shallow water. Preys on fish, but also takes sea-snakes, cephalopods and crusteaceans.

Carcharhinus sorrah (Valenciennes, 1839)

Spot-tail shark

Length: To 1.6 m.
Distribution: Widespread tropical Indo-Pacific, including all of our area.
Depth: 1-80 m.
General: An open water species occasionally near reefs. Prefers large open sand areas adjacent to reefs and usually found well above the substrate and in surface waters. For comparison it is about as common as the Blacktip reef shark by fisheries catches, but rarely seen. It has a distinc black tip on the lower lobe of the caudal fin.

REQUIEM SHARKS • CARCHARHINIDAE

Dusky shark

Length: To 3.6 m.
Distribution: Cosmopolitan in various tropical and temperate areas. In our area mainly along contental margin of China and Vietnam.
Depth: 1-400 m.
General: A locally abundant species, sometimes forming large schools in coastal estuaries.
They feed on a variety of fish and cephalopods throughout the water colomn but are usually seen swimming near the bottom. A slender species, dark grey over the top and white below.
In Australia often called black whaler and sometimes bronze whaler, however the latter belongs to C. brachyurus and has led to confusion of its range.
The latter is a temperate open water temperate species which doubtfully occurs in our area.

Carcharhinus obscurus (Lesueur, 1818)

Sandbar shark
Length: To 2.4 m.
Distribution: Circumtropical, ranging to subtropical zones, appears to be confined to continental China and Vietnam in our area.
Depth: 1-280 m.
General: A deep water species, occasionally enters shallow depths including intertidal, but adjacent to deep water.
A timid species in the shallows as adult, but juveniles school in coastal areas used as nursery grounds. Feeds near the bottom on a variety of fishes, cephalopods and crustaceans. Best recognised by the tall dorsal fin, combined with plain pale colour.

Carcharhinus plumbeus (Nardo, 1827)

REQUIEM SHARKS • CARCHARHINIDAE

Carcharias falciformis (Bibron, 1839)

Silky shark

Length: To 3.3 m-
Distribution: Circumtropical, in our area known from Vietnam, Philippines and New Guinea.
Depth: 1-500 m.
General: This species prefers water above 23 degrees Celsius. Diet consists mainly of bony fishes, but also includes cephalopods and pelagic crabs. It is often associated with schools of tuna.
Distinct feature is the first dorsal fin originating behind the pectoral fin tips.

Carcharhinus leucas (Valenciennes, 1839)

Bull shark

Length: To 3.4 m.
Distribution: Circumtropical in most seas, throughout our area.
Depth: 1-150 m.
General: A large heavily built plain species, dusky above and pale below. Coastal areas from deep flat sand and rubble areas near reefs to freshwater rivers. Reported from nearly 4000 km from the sea in the Amazon. A dangerous shark in coastal rivers, responsible for numerous attacks on swimmers in Australia and Africa, and also taken swimming dogs on several occasions.

Galeocerdo cuvier (Peron & Lesueur, 1822)

Tiger shark

Length: To 5.5 m.
Distribution: Circumtropical, including all of our area.
Depth: 1-300 m.
General: The most dangerous shark in the tropics. A scavenger eating virtually everything in the sea and terrestrials taking to the water. Preys on turtles during the nesting periods in many areas. Occurs mainly along outer reefs, and consequently is not commonly seen by divers, however it can be expected anywhere. This species will feed on dead animals and their smell could bring tiger sharks into coastal waters.

REQUIEM SHARKS • CARCHARHINIDAE

Oceanic whitetip shark

Length: To 3.5 m.
Distribution: Circumtropical, throughout our area.
Depth: 1-150 m.
General: Mainly a pelagic species, easily recognised by the large fins which broadly tipped white. Often seen from boats as it slowly swims just below the surface, usually accompanied by pilot fish or other sharks. This gracious swimmer is only occasionally in close vicinity of reefs. Reported as one of the four most dangerous to humans but no attacks to divers are known.

Carcharhinus longimanus (Poey, 1861)

White-tip reef shark

Length: To 2.1 m.
Distribution: Indo-Pacific, including all of our area, and east Pacific.
Depth: 5-300 m.
General: Probably the most observed shark by divers. Occurs commonly on coral reefs but usually resting under plate corals or deep on sand flats where they often congregate in small groups. Occasionally swims along reefs during the day, unaffraid of divers and can be approach at close range in some areas. A distinct species with the white tipped dorsal and caudal fins, combined with slender body.

Triaenodon obesus (Rüppell,1837)

Blue shark
L: To 3.8 m. Di: One of the most widespread sharks found in all large open seas.
De: 20-220 m. Ge: An open sea pelagic over continental shelf, rarely seen close to reefs, except where shelf is narrow. Prefers temperatures of 12-20 degrees Celsius and in the tropical it occurs very deep or in cold upwellings such as in southern Indonesia. They are fast swimmers and may travel over a great distance, a specimen tagged off Tasmania was recaptured south of Java. In interesting fact is the skin of females is twice as thick as that of males to cope with the male's courtship which must be a rough affair.

Prionace glauca (Linnaeus, 1758)

21

REQUIEM SHARKS • CARCHARHINIDAE

Grey reef shark
Length: To 1.8 m.
Distribution: Indo-Pacific, including all of our area.
Depth: 10-280 m.
General: Probably the most numerous shark on coral reefs but shy and timid in areas not used to divers. Quickly becomes accustomed to divers feeding them but can become aggressive under baiting conditions. Usually forms groups, working together when hunting fishes. This species is mainly grey, sometimes with a brownish tinge dorsally, an a broad pale indistinct band from over the gill slits along belly. A narrow pale margin on dorsal fin tip and a dark posterior margin on caudal fin broadening over lower lobe.

Carcharhinus amblyrhynchos (Bleeker, 1856)

HAMMERHEAD SHARKS • SPHYRNIDAE

Scalloped hammerhead

Length: To 3.5 m.
Distribution: Circumtropical, including all of our area.
Depth: 1-275 m.
General: This is the common hammerhead in tropical waters, usually seen solitair, but form large migratory schools. Mainly pelagic but often swim near reefs over the deep drop-offs. Adult females are rare inshore and seem to stay more in deep water where they mate and give birth.
They feed on a variety of fishes and cephalopods.

Sphyrna lewini (Griffith & Smith, 1834)

HAMMERHEAD SHARKS • SPHYRNIDAE

Great hammerhead

Length: To 6 m.
Distribution: Circumtropical, including all of our area.
Depth: surface to 80+ m.
General: Pelagic over continental shelf, but entering very shallow depths in coastal bays at times. This large species is probably responsible for troubling spearfishermen. Feeds primarily on demersal prey, including crustaceans, cephalopos and fishes of various kinds. Apart from size, this species is best recognised by the tall dorsal fin with falcate shape.

Sphyrna mokarran (Rüppell, 1837)

Smooth hammerhead

Length: To 4 m.
Distribution: Globally in temperate and subtropical regions, in our area ranging along mainland from the north to Gulf of Thailand.
Depth: 1-20 m.
General: Swims above continental shelves in surface waters, occasionally inshore. Feeds primarily on squid and some fishes, but diet varies between areas and may include benthic crustaceans or elasmobranch fishes. Because of pelagic nature usually seen solitair, but may occur in large schools off-shore.

Sphyrna zygaena (Linnaeus, 1758)

ANGEL SHARKS • SQUATINIDAE

Japanese angel shark
Length: To 2 m.
Distribution: Tropical Japan to Philippines.
Depth: 20-100+ m.
General: Burries in the sand during the day to ambush prey, but may go on the prowl for food at various times.
Feeds on a variety of sand dwelling invertebrates such as octopus or cuttlefish. Found on slightly sloping sand zones near reefs, strategic spots for likely prey traffic. Angel sharks have a distinct shape and sharp teeth. They are capable of biting their own tail and should not be handled.

Squatina japonica Bleeker, 1854

SAWFISHES • PRISTIDAE

Large fishes, growing the longest of all living rays, some reaching a length of over 6 m. The armed sawfishes live in marine as well as freshwater habitats, some travelling a long way up rivers. Only four marine species are known worldwide. They are distinguished from the superficially similar saw-sharks with two obvious features. The saw-sharks have gill slits on sides of head which are underneath in sawfishes (like other rays) and the rostral teeth generate in saw-sharks but not in sawfishes. The „saw" is studded with about 30 pairs of lateral teeth and is used to catch prey by slashing the rostrum sideways through dense schools of small fishes. They also dig in the bottom like other rays by blowing a jet of water from the mouth through the spiracles on top to expose prey such as molluscs and crustaceans. The toothed rostrum is also effective as self-defense as injuries found on captured sharks suggests. Gravid females are protected from the fetus which have a protective membrane over the rostrum which in addition is flexible rather than rigid soon after birth.

Narrow sawfish
Length: To 3.5 m.
Distribution: Widespread tropical Indo-Pacific, including all of our area.
Depth: 2-40 m.
General: Reported much larger, but probably based on other species which can attain 7 m, some of which living in pure freshwater. Usually in muddy coastal bays, estuaries and mangrove habitats. The various species are identified by the teeth on the rostrum and position of the first dorsal fin in relation to pelvic fins. This species has about 20 rostral teeth on each side but they are concentrated over the frontal half.

Anoxypristis cuspidata (Latham, 1794)

Wide sawfish
L: To 6.5 m.
Di: Probably widespread, in our area known only from Philippines. De: 5-40 m.
Ge: The largest of the sawfishes.
Feeds on slow swimming fishes and digs for invertebrates in the sand. Mainly in estuaries and mangrove swamps.

Pristis pectinata (Latham, 1794)

SHARKFIN GUITARFISHES • RHYNCHOBATIDAE

White-spotted guitarfish

Length: To 3 m.
Distribution: Indo-Pacific, including all of our area.
Depth: 1-50 m.
General: A large stocky ray which is shark-like posteriorly and can attain a weight of more than 220 kg.
Common in some areas on shallow sand flats and is inquisitive, often coming to investigate divers. Highly variable, probably related to habitat, from pale yellowish brown to totally black, usually with some white or pale spotting and a dark ear-like spot. Possibly more than one species under this name.

Rhynchobatus djiddensis (Forsskal, 1775)

Shark ray

Length: To 2.7 m.
Distribution: Indo-Pacific, including all of our area.
Depth: 3-90 m.
General: Probably the most unusual ray we can encounter. Looks more like a shark and swims like it as well. Found in open sand areas near or along base of reefs. Easily recognised by its deep body, broadly rounded snout and horny ridge with large thorns on back.

Rhina anclystoma Bloch & Schneider, 1801

TORPEDO RAYS • TORPEDINIDAE

Marbled torpedo ray

Length: To 1 m.
Distribution: Indian Ocean, to Andaman Sea.
Depth: 0.5-200 m.
General: Lagoon reefs around coral bommies and rocks. Aggregates in schools during the mating season. Animals stranded by low tide can survive for several hours.

Torpedo sinuspersici Olfers, 1831

STINGRAYS • DASYATIDIDAE

Stingrays are amongst the largest of cartilaginous fishes, with some species exceeding 2 m in disc width and a weight of 350 kg. The disc is variably depressed and the shape also variable, depending on the species, from oval, almost circular, to diamond-like. The head is only slightly raised above the pectoral fins and the is mouth armed with numerous small teeth. Denticles, thorns and tubercles, if present are found only on the dorsal surface of the body and tail. The latter is often much longer than the disc and whip-like without fins. It may have membranous skin folds which may look fin-like and many species possess one or more venomous serrated stinging spines.

Cowtail stingray

Width: To 1.8 m and length to 3 m with tail.
Distribution: Indo-Pacific, including all of our area.
Depth: 1-60 m.
General: A large uniformly dark ray but easily recognised by the deep skinfold below the tail. The sting is just above the beginning of the skinfold about halfway the tail if in tact. An active species, roaming the shallow coastal bays and enters estuaries. Reputed to investigate spearfishermen during their activities. Usually accompanied by piloting fish or remoras.

Pastinachus sephen (Forsskal, 1775)

STINGRAYS • DASYATIDIDAE

Jenkins whipray

Width: To 1 m, length to 2 m.
Distribution: Andaman Sea to Malaysia and Thailand.
Depth: 5-55 m.
General: A medium large yellowish brown stingray with a whip-like tail without skin folds. A row of enlarged spear-shaped thorns and a narrow band of close-set denticles extending along head, back and tail. Ventral surface, including disc-margin white.
Tail variably ending greyish. Seen singly and sometimes in groups.

Himantura jenkinsii (Annandale, 1909)

Pink whipray

Width: To 1.5 m.
Distribution: Indo-Pacific, Andaman Sea to Philippines, probably widespread.
Depth: 20-200 m.
General: A large species with very long tail, ist length may exceed 5 m. Plain uniformly brownish pink above.
An off-shore species, usually around atolls in deep water and probably have a taste for prawns as they are often caught by the trawlers targetting for these.

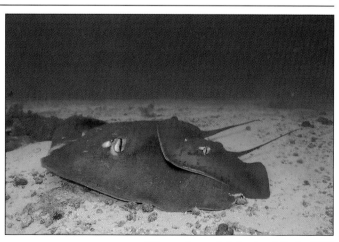

Himantura fai Jordan & Seale, 1906

27

STINGRAYS • DASYATIDIDAE

Mangrove whiptail

Wi: To 1 m. Di: In our area from Java, Indonesia, Andaman Sea and to Philippines. De: 05-10 m. Ge: Shallow sand flats in sheltered coastal bays, estuaries and around mangrove habitats. Best recognised by the white tail which appears as being dipped in white paint.

Himantura granulata (Macleay, 1883)

Leopard whipray

Width: To 1.4 m.
Distribution: Tropical west Pacific and east Indian Ocean, including all of our area.
Depth: 3-80 m.
General: A moderately common species, best recognised by close set variously shaped spots, often with light centres in larger individuals. Usually referred to as *H. uarnak,* our next species, also in our area which has a reticulated pattern looking like a large maze, and has a more rounded disc. Lives mainly on the inner continental shelf, occasionally entering shallow depths.

Himantura undulata (Bleeker, 1852)

Reticulated whipray

Width: To 1.5 m, length to 4.5 m.
Distribution: Indo-Pacific, throughout our area.
Depth: 3-175 m.
General: Tail longer than disc. Upper surface of disc and tail highly ornamented in adults, pale to yellow with dense dark-brown reticulations.
Common on continental shelfs and name used for several other similar species.

Himantura uarnak (Forsskal, 1775)

STINGRAYS • DASYATIDIDAE

Blue-spotted mask-ray
Wi: To 40 cm. Di: Tropical Indo-Pacific. De: 6-90 m. Ge: A common species on coastal sand and rubble slopes, usually well away from reef. Often buried during the day with just eyes exposed. Has small ventral fold on tail with sting above just in front. Easily recognised by the long tail with white tip.

Dasyatis kuhlii (Müller & Henle, 1841)

Blue-spotted fantail ray

Width: To 30 cm.
Distribution: Widespread tropical Indo-Pacific, common throughout our area.
Depth: 2-30 m.
General: The most commonly observed ray on coral reefs. During the day usually under table corals or in ledges, becoming active on dusk, moving along the reefs over sand patches.
In some areas responds to tides, feeding on the shallow flats on high tides. A small species with about 70 cm maximum including tail. Sting well back on tail and often has two.

Taeniura lymna (Forsskal, 1775)

29

STINGRAYS • DASYATIDIDAE

Blotched fantail ray

Width: To 1.8 m.
Distribution: Widespread tropical Indo-Pacific, including our area.
Depth: 6-100+ m.
General: A large ray, reaching 3 m if tail intact. Highly variable from pale grey to almost black with irregular darker blotches. Usually has one sting at distance from tailbase. Usually seen along the base of drop-offs or on sand flats near reefs. Often digs large holes by blowing sand from the mouth, taken in by the large spiracles on top, to dislodge molluscs and crabs. Un apparently undescribed Himantura is shown in the photograph also, taken in New Guinea. Usually identified as *T. melanospilos*, a synonym.

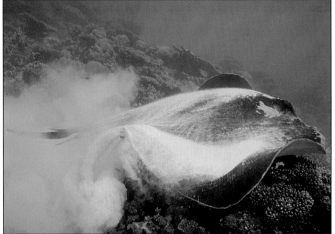

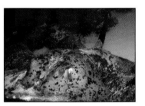

Taeniura meyeni Müller & Henle, 1841

Porcupine ray
Wi: To 1 m. Di: Tropical Indo-Pacific, including all of our area, and east Atlantic. De: 1-30 m. Ge: A distinct species by its oval and very rounded outline, and thorn-like spines all over back. Has no stinging spine. Shallow sand flats, usually near seagrass beds or rubble reefs in coastal and inner reef areas.

Urogymnus asperrimus (Bloch & Schneider, 1801)

STINGAREES • UROLOPHIDAE

Sepia stingaree
Width: To 30 cm.
Distribution: North-west Pacific from southern Japan to East China sea, Vietnam, in our area.
Depth: 10-200 m.
General: Differs from stingrays in having a distinct caudal fin. Venomous spine on tail large and is used readily if disturbed, striking like lightning. Can pierce deeply and very painful, causing prolific bleeding. Application of heat with water or blowing with hairdrier alleviates pain. A small species found on sand adjacent to rocky reefs. Often buries during the day near or under ledges.

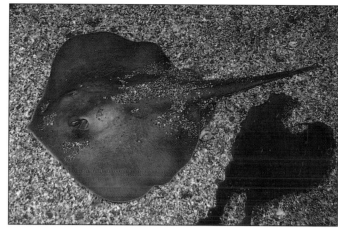

Urolophus aurantiacus Müller & Henle, 1841

COWNOSE RAYS • RHINOPTERIDAE

Java cownose ray

Width: To 1.5 m.
Distribution: Indo-Pacific, including all of our area.
Depth: 1-40 m.
General: A stocky ray with a lozenge-shaped disk, a whip-like tail and a distinctive bilobed forehead with a pair of lobe-like flabs located underneath. Mainly coastal, on mud-flats, in estuaries and sandy shallow bays. Usually in schools often with hundreds of individuals. An active and fast swimmer, occasionally stopping to feed on substrate, taking molluscs and fish.

Rhinoptera javanica Müller & Henle, 1841

EAGLE RAYS • MYLIOBATIDIDAE

White-spotted eagle ray

Width: To 3 m, rare over 2 m.
Distribution: Circumtropical, throughout our area.
Depth: 1-40+ m.
General: Magnificent large ray with very long tail, reported total length of disc and tail is almost 9 m. Easily recognised from the top with the numerous small white spots. Mainly in open water, often travelling in large schools near the surface or over the bottom of vast flat sand streches between shore and outer reefs. It often make great leaps, jumping clear from the surface. Feeds primarily on molluscs (left below).
The copulation of eagle rays has been described: The ♂ gets a tight hold of the ♀ by biting into the pectoral fin and subsequently slips under her, belly to belly, to insert his claspers. This takes place either in the open water while swimming, or on the bottom and may last 30 to 90 seconds. A ♀ may copulate with several males in succession. Ovo-viviparous. After a gestation period of 12 months gives birth to 1 to 4 young in shallow coastal waters, which are already 50 cm long at the time of birth. Eagle rays are an important prey for pelagic sharks. Other sharks follow the eagle rays into the shallow bays to feed on the newly born. Although they are mostly observed swimming, eagle rays do not feed in the open water. Rather, they dig for molluscs, worms, crustaceans, and cephalopods in coral rubble in the late afternoon. Reef fishes such as goat fishes and wrasses are attracted by this activity and follow the eagle rays as commensals profiting from the small invertebrates which are disturbed by the rays. Unlike manta rays, eagle rays are rarely "cleaned" by cleaner wrasses near reef areas.

Aetobatus narinari (Euphrasen, 1790)

DEVIL RAYS • MOBULIDAE

Pygmy devil ray

Width: To 1 m.
Distribution: Widespread tropical Indo-Pacific, including all of our area.
Depth: 1-20 m.
General: The smallest devil ray, usually occurs off shore in groups. In our area mainly seen in current areas such as in the Komodo islands with great tidal acticity. It has a large head compared to the other species, most similar is M. *javanica* which grows much larger and usually has a small stinging spine at tail base.
Another similar species in our area is the sicklefin devil ray M. *thurstoni* which has a concave anterior margin towards the apex of the disc.

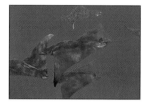

Mobula eregoodootenkee (Cuvier, 1829)

Manta ray
Wi: To 6.7 m, reported to 9 m.
Di: Circumtropical. De: 1-40 m.
Ge: The largest ray, pelagic, but commonly observed near reefs in pursuit of plankton which are guided to the enormous mouth by broad fleshy lobes. These forward projecting parts are an extention of the pectoral fins. They often feed near the surface when plankton has accumulated from currents along the reefs.

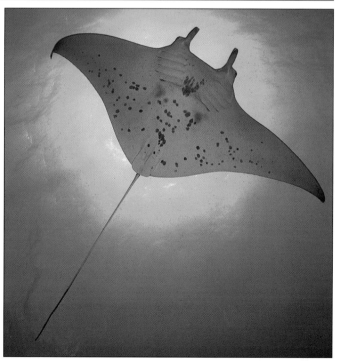

Manta birostris (Dondorff, 1798)

MORAY EELS • MURAENIDAE

A large family of marine eels, comprising at least 10 genera and well over 100 species (200 by some authors), distributed world-wide in all tropical and subtropical seas. They feature muscular robust posteriorly compressed bodies, with tough thick skin and a single median fin present, usually originating dorsally before gill opening, continuing around tail to anus. Mouth up to moderately large size, with variable teeth between genera. Gill openings reduced, as a small hole, mid-laterally behind head. Most species have distinct colouration, consisting of spots and stripes, often with complicated patterns which are usually typical for a species. Mostly reef dwellers, living secretive in holes and crevices. Usually observed with only the head exposed, however a few species move about on reefs in the open hunting for food. They feed on a variety of invertebrates and some species feed on fish. Although generally not aggressive, the larger species can inflict serious and painfull wounds. Some have relatively poor eye-sight and rely more on smell, often coming towards a diver to investigate. Sometimes they congrate in certain areas apparently to spawn, and in such areas they are known to attack divers.

Our knowledge of breeding behaviour is limited to a few species seen intertwine and release pelagic eggs. Some eels are known to migrate over great distances but most tropical reef eels remain in their home territory or congregate in small groups in some areas. Larvae have deep greatly compressed and clear bodies, settling at about 50 mm length as they change shape to the more rounded as in the adults. As adult they range in size from about 30 cm to 2 m.

Identification of the common and during the day more obvious species is not difficult. Mpst feature unique head or body patterns, or a certain spot or colour in a particular place. However, there are a number of drab plain brown and grey species which are secretive and usually only seen at night, and these can be difficult. Identification of just pictures is almost impossible, and may only get identified to the genus. Even identifying the genus can be difficult as it is mostly related to the dentition and shape of jaws. The genus *Siderea* maybe difficult to separate from *Gymnothorax* because it mainly differs in type of teeth in the jaws.

Ribbon eel

Length: To 1.2 m.
Distribution: Indo-Pacific, including all of our area.
Depth: 1-55 m.
General: The most distinct eel on tropical reefs. Usually the head and part of the body may stretch out in trying to grab a small fish as prey. Juveniles are black and adults blue with variable amount of yellow. Coastal reef crests and slopes, usually seen singly but occasionally a pair may share the same burrow.

Rhinomuraena quaesita Garmann, 1888

MORAY EELS • MURAENIDAE

Specimen swimming with diver shows the great length of the body.
Ribbon eels area rarely seen leaving their burrow.

Below a large black phase specimen about to dissapear and a juvenile reaching out to a damselfish.

Rhinomuraena quaesita Garmann, 1888

Honeycomb moray

Length: To 2 m.
Distribution: Throughout the tropical Indo-Pacific, including all of our area.
Depth: 3-50 m.
General: One of the largest morays, easily identified by the colouration. Coastal protected shallow bays to deep reefs. Rare in the more eastern parts of the west Pacific.

Gymnothorax favagineus (Bloch & Schneider, 1801)

MORAY EELS • MURAENIDAE

Dragon moray

Length: To 80 cm.
Distribution: Indo-Pacific, including all of our area.
Depth: 15-50 m.
General: A rare species in our area and only commonly observed in southern Japan and northern Philippines, but reported to northern Australia and various locations in the Indian Ocean. Appears to prefer rocky, rather than coral dominated reefs.

Enchelycore pardalis (Temminck & Schlegel, 1843)

Barred moray

Length: To 60 cm.
Distribution: Indo-Pacific, including all of our area.
Depth: 1-15 m.
General: A shallow reef species found on clear coastal flats and lagoons. Secretive, hunting under the cover of the reef in search for small fish or shrimps. Opportunity feeder, but mostly observed at night because shades during the day is the cover for this species.

Echidna polyzona (Richardson, 1844)

Clouded moray

Length: To 70 m.
Distribution: Indo-Pacific, including all of our area.
Depth: 1-15 m.
General: Coastal and very shallow reef flats from the intertidal zone to lagoons on inner reefs. Sometimes found under rocks in pools at low tide. May hunt in bright sunlight in shallow lagoons, moving about in the open over rubble patches. Also known as Snowflake moray.

Echidna nebulosa (Ahl, 1789)

MORAY EELS • MURAENIDAE

Golden moray

Length: To 26 cm.
Distribution: Indian Ocean to Andaman Sea and Marshall Islands.
Depth: 5-25 m.
General: A small rare eel known from a few locations. With the bright colour it should be easily seen if in the area. Coastal bays on sponge and coral reef.

Gymnorhorax melatremus Schultz, 1953

Yellow-edged moray
L: To 1.2 m. Di: Indo-Pacific, including all of our area. De: 3-100+ m. Ge: A common species on most reefs, especially mixed rock and corals. They participate quickly in shark or cod feeding areas and have caused nasty wounds to inexperienced divers. Juveniles below.

Gymnothorax flavimarginatus (Rüppell, 1828)

Giant moray

Length: To 2.4 m.
Distribution: Indo-Pacific, throughout our area.
Depth: 10-50 m.
General: One of the largest morays which has potential to inflict injury to divers if not careful when feeding other fish, practised in some tourist operations. Found in various reef habitat from coastal to outer reefs.

Gymnothorax javanicus (Bleeker, 1828)

MORAY EELS • MURAENIDAE

Gymnothorax undulatus (Lacépède, 1803)

Undulate moray

Length: To 1 m (reported to 1.5 m.).
Distribution: Indo-Pacific, including all of our area, ranging from east Africa to Panama.
Depth: 5-50 m.
General: A common species, mostly observed at night when it hunts by moving around in search for small fish or shrimps. In Bali this species was photographed taking a cardinalfish from a hole in a sponge. Coastal reef slopes, can be aggressive towards divers.

Gymnothorax zonipectus Seale, 1906

Bar-tail moray

Length: To 46 cm.
Distribution: Indo-Pacific, throughout our area.
Depth: 6-30 m.
General: This species seems to like rich coral growth reefs where it hides during the day and moves about more in the open at night.
A distinct species but only by the patterns of the more posterior part of the body.

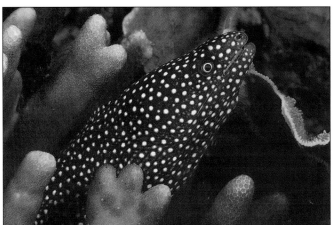

Gymnothorax meleagris (Shaw & Nodder, 1795)

White-mouth moray

Length: To 1 m.
Distribution: Indo-Pacific, throughout our area.
Depth: 1-35 m.
General: Uncommon in our area, mainly from Manado/Sulawesi north to the Philippines. north. Clear coastal and protected inner reefs, usually living amongst corals in rich coral-algae mixed habitats. Best recognised from similar species by the white inside of mouth.

MORAY EELS • MURAENIDAE

Spot-face moray

Length: To 40 cm (reported much larger).
Distribution: Indonesia to Queensland, and to Philippines, southern Japan, and Micronesia.
Depth: 10-50 m.
General: Only known commonly from Bali, reported as rare elsewhere. At Bali it seems to like the steep sand slopes with many but isolated outcrops of reefpatches and sponges in which they make their home, especially where shrimps have well established cleaning stations.

Gymnothorax fimbriatus (Bennett, 1831)

White-eyed moray

Length: To 65 cm.
Distribution: Indo-Pacific, throughout our area.
Depth: 1-35 m.
General: Coastal, often silty habitat, reefs, lagoons and often on isolated reef outcrops on mud or sand slopes. A distinct species, best identified by the white eyes. It is common throughout our area and often found in pairs or mixed with another species, especially the next species.

Siderea thyrsoidea (Richardson, 1845)

Zebra moray

Length: To 1.5 m.
Distribution: Indo-Pacific, throughout our area.
Depth: 2-50 m
General: Coastal to outer reef crests and slopes. Very secretive, virtually only seen at night when they hunt the reefs for small fishes and shrimps. Easily idenfied with the distinct banding.

Gymnomuraena zebra (Shaw, 1797)

SNAKE EELS • OPHICHTHIDAE

A very large family of marine eels, comprising over 50 genera and 250 species worldwide. Highly diverse, moderately robust to extremely long tubular bodies. Head usually sharply pointed and often ending in a strong pointed tail. Various colour patterns from uniformly pale to spotted, or conspicuously banded. Usually they are buried in sand or mud and rarely sighted during the day. A few banded species move about in the open, usually in shallow coastal areas, resembling sea snakes. Most species feed on fishes and crustaceans. Size of adults varies greatly between species, raning from about 20 cm to 2.5 m.

There are a great number of species in south-east Asian waters, but because of their secretive habits even the most common species are rarely seen. The most seen of the burying species is the head and this usually is camouflaged Included here are the two only species occasionally seen out in the open as they hunt for small fishes and crustaceans.

Banded snake eel

Length: To 97 cm.
Distribution: Indo-Pacific, throughout our area.
Depth: 1-25 m.
General: This species is often seen by snorkelers and usually mistaken for a sea snake. It mimics the venomous sea snakes in many areas and the colour pattern varies accordingly. In Japan a banded sea snake eating the banded snake eel was photographed. Pacific form fully banded but Indian Ocean populations are often half banded or have round spots alternating with bands.

Myrichthys colubrinus (Boddaert, 1781)

Spotted snake eel

Length: To 99 cm.
Distribution: Indo-Pacific, throughout our area.
Depth: 1-30 m.
General: Highly variable in size and number of spots, changing from few and large to many and small in adults. Shallow coastal sand flats, reef crest, drop-offs and lagoons, mainly out hunting at night.

Myrichthys maculosus (Cuvier, 1816)

GARDEN EELS • HETEROCONGRIDAE

The garden eels are closely related to the large conger-eel group, the latter mostly found deep, in muddy habitats, or in temperate zone, and are rarely enclountered on or near reefs. The garden eels however are an interesting group of fishes for the scuba diver, especially as the name applies they are found in such great numbers together that it looks like a garden of eels. The group is under revision and several species are in the process of being described.

The members of this family typically are seen on open sand patches or flats within or well away from reefs, with head upwards and part of the body vertically protruding from the bottom. Some species are extremely long and thin at the same time, reaching for above the substrate to feed on zooplankton drifting accross. Usually these occur in dense and countless numbers, sometimes in a field with seemingly no end. Most species live in current prone areas, often running strong with certain tides, and most species feed when the currents bring the food either from outflowing lagoons or incoming oceanic waters, depending on the type of plankton living there.

Black garden eel

Length: To 60 cm.
Distribution: Indonesia and Philippines
Depth: 1-35 m.
General: A common species on quiet Indonesian coasts from very shallow sandy or muddy flats to moderately deep flats. In the shallows mostly in small groups but deeper and where sand flats strech a long way from shore they may congregate in 'garden-like' numbers. They appear black underwater but they vary from dusky to chocolate-brown.

Heteroconger perissodon Böhlke & Randall, 1981

Spotted garden eel

Length: To 40 cm.
Distribution: Indo-Pacific as far as Africa, Red Sea and Samoa, including all of our area.
Depth: 6-45 m.
General: Found mainly on large open sand patches between reefs and on gradual slopes. Usually in depths of 15 m or more but occasionally patches are found very shallow depths where there is protection and currents providing food. Photograph was taken in Bali at only 6 m depth. Divers notice this species most because it lives near reefs.

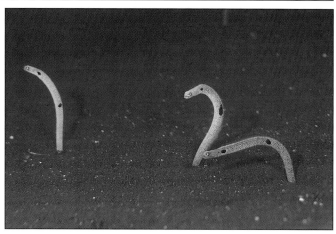

Heteroconger hassi (Klausewitz & Eibl-Eibesfeldt, 1959)

HERRINGS • CLUPEIDAE

A large family with 65 genera and 180 species distributed world-wide, constituting an important food source for predators, thus an important link in the food chain. People also take a big bite out of the populations as they are the greatest single group exploited by fisheries and this has led to wiping out some species or population, which in turn had upset the balance of the local fauna, especially effecting seabirds. Apart from herring the family includes the pilchards and sardines. They school in great numbers inshore and off shore, swimming so dense that they appear at distance as a single body. Most species are small, the largest about 25 cm long. They feature a single centrally placed dorsal fin and lack a distinct lateral line. Other features are a large mouth with distensible jaws and the presence of numerous and long gill rakers are a typical feature for plankton filter feeders.

From a divers point of few this family is seen as a school of silvery fishes and indeed the various species are difficult to identify without having a specimen in hand. Included a common and widespread species.

Goldspot herring

Length: To 14 cm.
Distribution: Indo-Pacific from Africa to central Pacific, including all of our area.
Depth: 1-40 m.
General: Protected coastal bays, lagoons on large reefs or islands, and swimming in densely packed schools over sand flats or along slopes. The similar but much smaler species are known as sprats or sardines.

Herklotsichthys quadrimaculatus (Rüppell, 1837)

ANCHOVIES • ENGRAULIDAE

A family of small, to about 20 cm, silvery fishes with 16 genera and 140 species world-wide. Like the closely related herrings they form an important link in the food chain, preyed on by pelagics such as tune and many sea birds, and also for man. The prominant snout protrudes well over the mouth, a feature which easily identifies these fishes when seeing them underwater. The mouth opens wide and feeding fishes swim fast with the mouth fully extended to filted plankton from the water with their numerous long slender gill rakers.

Anchovies are mostly coastal fishes in shallow depths, though sometimes found to 200 m, and commonly enter estuaries and tidal channels. A large number of species occurs in our area but they are very similar in general appearance and difficult to identify without examining a specimen. Included here one of the common species often seen near coastal reefs or under jetties.

ANCHOVIES • ENGRAULIDAE

Little priest

Length: To 15 cm.
Distribution: Indo-Pacific from Red Sea to Samoa, including all of our area.
Depth: 1-200 m.
General: Forms large and dense schools in coastal bays, often congregating near freshwater runoffs. Quick swimming fishes and darting all over the place in feeding frenzies when encountering a cloud of plankton.

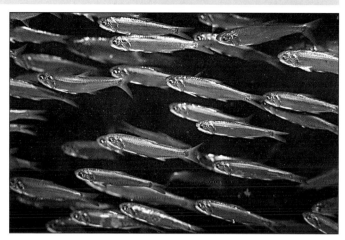

Thryssa baelama (Forsskal, 1775)

EELTAIL CATFISHES • PLOTOSIDAE

There are two main types of catfish which is most obvious by the difference in the tail. The eel-like tail in which the dorsal, anal and caudal fins are connected into one confluent fin, and the fork-tailed with a separate forked caudal fin. There are more than 2000 species of catfish but most occur in freshwater or turbit coastal estuaries and river mouths. Only one species of eeltail catfish is commonly observed on reefs in tropical waters which is included here.

Avoid handling these fishes as the fin spines are highly venomous. Stings are extremely painful and repeated stings can be fatal. If stung apply heat (water or air-blower as hot as one can bare) to the wound as soon as possible.

Striped catfish
L; To 35 cm. Di: Indo-Pacific, including all of our area.
De: 1-50+ m. Ge: Juveniles are commonly seen out in the open, feeding on the sand or mud in dense groups. The smaller the juvenile the denser the schooling. Mostly coastal estuarine habitats where you find land-debris such as sticks, coconuts etc.

Plotosus lineatus (Thunberg, 1787)

LIZARDFISHES • SYNODONTIDAE

A primarily tropical family, comprising two genera and about 35 species worldwide. Most are reef predators, featuring a torpedo shaped body, a large mouth with long needle-like teeth along entire jaws to seize prey. On tropical reefs they typically rest on top of reefs, strategically chosen, to view the are for possible prey. Camouflaged and waiting motionless in ambush, with lightning speed a careless fish is taken which are usually mostly juveniles of the larger reef dwellers.

Most species are found on or near reefs, some of hard substrate but usually on sand patches on reef crests or rubble slopes, sometimes burying in the substrate. They can be found coastal as well as on outer reefs, depending on the species. Identification of some species is difficult with a few distinct features available and great similarities in general appearance. In addition there is strong variation within each species as with most camouflage species, matching surroundings or a particular habitat.

Red-marbled lizardfish

Length: To 12 cm.
Distribution: Indonesia, Malaysia, Taiwan, Philippines and northern Australia.
Depth: 10-50 m.
General: A small species, easily overlooked. Prefers clear protected reefs, usually along base of drop-offs on rubble, rock or corals.
The banding is slightly more defined and the snout pointed compared to similar species.

Synodus rubromarmoratus Russell & Cressey, 1979

Grey-streak lizardfish

Length: To 22 cm.
Distribution: Indo-Pacific, including all of our area.
Depth: 1-50 m.
General: Coastal sand flats and slopes to outer reef lagoons. Mostly found on sand near reefs and often buries in the sand, single or small groups when males compete to spawn with gravid female. Best identified by the grey streak along sides at eye-level.

Synodus dermatogenys Fowler, 1912

LIZARDFISHES • SYNODONTIDAE

Variegated lizardfish

Length: To 25 cm.
Distribution: Indo-Pacific, throughout our area.
Depth: 3-50 m.
General: By far the most common reef dwelling lizardfish. Mostly resting on hard substrate, including live coral and sponges, or immediate sand slope, singly or in pairs. Coastal to outer reefs. Variable colour, banding from dull grey to bright red.

Synodus variegatus (Lacépède, 1803)

Tail-blotch lizardfish
L: To 20 cm. Di: Indo-Pacific. De: 3-50+ m. Ge: Coastal to off-shore, mostly on sand and only occasionally on or near reefs. Often in quiet parts of large sandy lagoons where they bury near the upper edge of sand slopes. One of the few species easily identified by the black blotch on the tail.

Synodus jaculum Russell & Cressey, 1979

Painted lizardfish

Length: To 30 cm.
Distribution: Indo-Pacific, including all of our area, and Atlantic.
Depth: 0.5-200+ m.
General: Protected coastal and estuarine sand flats from subtidal zones to deep off shore habitats. Lives in spread out groups on sand flats and slopes and remains totally buried with only just the eyes exposed. Only when chasing each other or when disturbed they come out of the sand, quickly dashing away to bury as soon as stopped.

Trachinocephalus myops (Forster, 1801)

GRINNERS • HARPADONTIDAE

A small tropical family with two genera and about 15 species. They are closely related and comparable to the lizardfishes but have multiple bands of teeth, a spine in the ventral fins as obvious differences. As the lizardfishes they are voracious predators, taken rather large prey from ambush with great speed, however most are deep water dwellers and live on muddy or sandy substrates. They are also known as sauries and the trawled species are called bombay ducks. Few species are found on or near reefs and included are the more commonly encountered species in our area.

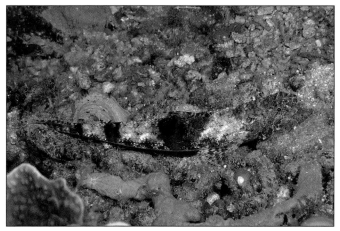

Saurida gracilis (Quoy & Gaimard, 1824)

Slender grinner

Length: To 28 cm.
Distribution: Indo-Pacific from Red Sea to Hawaii, including all of our area.
Depth: 3-100+ m.
General: The only species commonly found on rocky and coral reefs but found in all coastal, outer reef lagoons and deep off shore habitats. Highly variable, usually with two large black saddles on back below dorsal and andipose fins. Usually seen singly but sometimes in pairs, sometimes buries in sand.

Saurida nebulosa Valenciennes, 1849

Blotched grinner

Length: To 16 cm.
Distribution: Indo-Pacific from Mauritius to Hawaii, including all of our area.
Depth: 0.5-100+ m.
General: Shallow coastal, estuarine and Mangrove habitats but ranging to deep muddy off-shore habitats as well. Open mud and sand flats, sometimes near the edge of reef on rubble.

GRINNERS • HARPADONTIDAE

Long grinner

Length: To 25 cm.
Distribution: Probably Indo-Pacific, but only known from southern Japan, Philippines and Indonesia. Depth: 10-100+ m. General: Coastal and deep lagoon sand slopes, nearly always buried in substrate during the day. Out at night, suggesting feeding then and probably takes the nocturnal apogonids which drift near the bottom in these habitats.

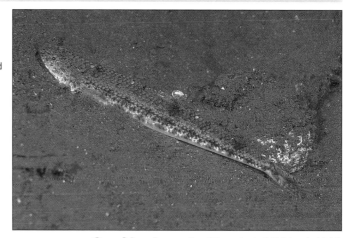

Saurida elongata (Temminck & Schlegel, 1846)

LANTERNFISHES • MYCTOPHIDAE

A large family of small benthopelagic fishes with some 250 species in 32 genera recognised. These fishes live diurnally in depths between 400-1000 m and one might be surprised to find this family represented in our book, however included here are underwater photographs taken on convential scuba diving equipment in Bali. Finding this species during a night dive caused great excitement, not only to the photographer but later to the scientific community and photographs from this occasion were published in several magazines. It was unusual to find these fishes so close to shore as until then only specimens were known from the open seas. The species undertake verticle migration at night and a few species reach near surface depths on dark nights to feed on zooplankton. They are a very interesting group of small species, the largest about 20 cm long but some mature at less than 30 mm, which have series of light-organs along the lower parts of the head and body. The lights function is a unexpected way of camouflage: the intensity matching the available light reflecting on the upper parts from the stars or other sources to eliminate the shadow.

Brooch lanternfish
L: To 80 mm. Di: Tropical west Pacific, Indonesia, Philippines. De: 25-400+ m. Ge: One of the few species in the family which can be seen in relativily shallow depths. Photographed in Bali at 25 m depth at Tulamben, north coast on a dark night (no moon). Picture below shows light-organs along underside.

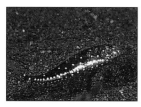

Benthosema fibulatum (Gilbert & Cramer, 1897)

LINGS • OPHIDIIDAE

A large worldwide family with nearly 50 general and over 150 species, most living on the bottom in very deep to extremely deep waters. Only a few are found on coral reefs which are mostly small and so secretive that they are only found when collections with chemicals are done. Only one species, included here, is often observed but mainly at night when it hunts.

The lings are eel-like with shorter bodies and large gill openings. The ventral fins are feeler like, often placed forward below the head, and barbels are often present around the mouth. The species with barbels superficially resemble catfishes, but only with the head. The barbels are arranged differenly around the mouth but the fins show the greatest differences between the families.

Bearded rockling

Length: To 60 cm.
Distribution: East Indian Ocean, Andaman Sea to Australia and Japan, containing all of our area.
Depth: 3-50+ m.
General: Coastal to protected inner reefs, rock and coral with lots of crevices. Hunts on the reef at night but quickly retreats when caught in a light. Diet consists of fishes and crustaceans. Shown from top left: In the back of a cave during daytime, juvenile at night, close-up of mouth and hunting at night.

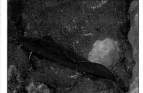

Brotula multibarbata Temminck & Schlegel, 1842

ANGLERFISHES • ANTENNARIDAE

This family are the shallow water members of a highly specialised groups which is unique in having a perculiar luring apparatus above the mouth, usually a modified first dorsal spine. This family has 12 genera and at least 41 species worldwide. They are mostly very cryptic reef dwellers which attract prey with their luring apparatus which is specialised in each species by mimicing the food-source of its prey. So the potential victim may see its favorite food, which in reality is a copy in looks and even motion, presented as a hard to resist easy meal but only to become a meal itself. The luring apparatus comes in two parts, the illicium or stork which varies greatly in length between species, and the esca with is the actual luring bit representing various types of crustaceans, worms or other fish. The lure is waved and moved to make the lure look like it's swimming.

Anglerfishes don't just position themselves anywhere but chose a spot where prey are likely to come along. The tropical species primarily feed on other fishes and with their expandable stomach can accomodate any large prey which passes through the mouth, including species longer then themselves. Many of the tropical species produce mucus egg rafts with numerous tiny eggs, but there are some which have demersal eggs and these are guarded by the female. Such eggs are much larger, up to 5 mm diameter, and much less numerous, some 5000 were counted in one species. Pelagic juveniles settle at about 25 mm but demersal egg hatchlings settle on the substrate and are about 10 mm long.

These interesting fishes are easily kept in an aquarium but introduction to community tanks are not recommended as the end result is one very healthy anglerfish. They need separate quarters and can be trained to take frozen food.

Sargassum anglerfish

Length: 13 cm.
Distribution: All tropical oceans but unknown from the east Pacific.
Depth: Surface (pelagic).
General: A very common species in floating weeds, especially sargassum weeds. May drift in to rockpools by wind driven currents. It was interesting to learn that when photographing this species in Flores in large weed rafts, this fish will jump on top of the weed, thus out of the water and lay there for considerable time before jumping back. A great way of getting away from a predator below.

Histrio histrio (Linneaus, 1758)

ANGLERFISHES • ANTENNARIIDAE

Shaggy anglerfish

Length: To 20 cm.
Distribution: Indo-Pacific from throughout our area, especially Indonesia and Malaysia.
Depth: 3-90 m.
General: Coastal bays with sponges, leaf litter on sand, hidden well with excellent camouflage and positioned in shaded areas. Variable from beige, light yellow, orange, yellow-brown to black. Usually with darker streaks or elongate blotches.

Antennarius hispidus (Bloch & Schneider, 1801)

Painted anglerfish

Length: To 16 cm.
Distribution: Indo-Pacific, including all of our area but severalgeographical forms.
Depth: 3-50 m.
General: A very common but one of the most variable species, virtually coming in all colours but blue and green, matching most sponge colours. A dificult species with local colour forms and perhaps there are a number of sub-species. Coastal reefs, lagoons and slopes with corals, sponges or at night just out in the open on sand or mud. Several colour forms shown.

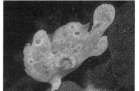

Antennarius pictus (Shaw & Nodder, 1794)

ANGLERFISHES • ANTENNARIDAE

Clown anglerfish

Length: To 10 cm.
Distribution: Indo-Pacific, including all of our area.
Depth: 1-15 m.
General: Small juveniles out in the open, looking very obvious but not as a fish but rather a seemingly poisonous nudibranch. Adults become more camouflaged. Coastal reefs in rocky areas. Typical for the family highly variable in colour, but juveniles normally bright yellow to white with orange to red markings.

Antennarius maculatus (Desjardins, 1840)

Giant anglerfish

Length: To 30 cm.
Distribution: Widespread tropical Indo-Pacific from Red Sea to Hawaii, throughout our area.
Depth: 1-50 m.
General: Shallow coastal to deep habitat but nearly always with sponges. Commonly found on jetty pilons which typically are situated in protected parts of islands. Comes in almost every colour except no reports of blue specimens.

Antennarius commersonii (Latreille, 1804)

CLINGFISHES • GOBIESOCIDAE

This large family comprises numerous tiny species, many of which only a few centimeters fully grown. About 35 genera and over 100 species are known but many species were only recently discovered and are without name, including several new genera. These fishes are particularly numerous in warm-temperate seas where corals are replaced by weeds.

Clingfishes are characterised by a large ventral sucking-disc evolved from united ventral fins as found in gobies. They lack scales and have a tough slimy coated skin instead, toxid in some species. They produce demersal eggs which are large and layed in small numbers on weeds or rock surfaces. The male or both sexes guard the eggs. Tropical species associate with featherstars (crinoids) and long-spined urchins. Included are the three commonly observed species in our area.

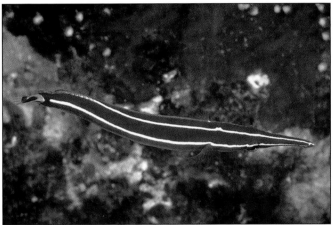

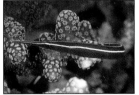

Long-snout clingfish

L: To 50 mm. Di: Indo-Pacific, including all of our area. De: 3-30 m. Ge: Swims openly near corals and is usually present when there are long spine urchins on the reef. Coastal reefs and protected inner reefs. Also known as striped clingfish but most other species have stripes.

Diademichthys lineatus (Sauvage, 1883)

Badger clingfish

Length: 75 mm.
Distribution: Andaman Sea.
Depth: 6 m.
General: Undetermined species, like several similar species which live under rocks or in ledges, and most are very small showing little colour. Many have a black stripe through the eye.

Lepidichthys sp

CLINGFISHES • GOBIESOCIDAE

One-stripe featherstar clingfish

Length: To 40 mm.
Distribution: Indo-Pacific, including all of our area.
Depth: 6-30 m.
General: A common species in our area, exclusively lives on featherstars (crinoids) and hide amongst the roll-up arms or underneath, usually living in pairs. Colour matches the host which can vary from yellow to reddish-brown or black.

Discotrema echinophila Briggs, 1976

Two-stripe featherstar clingfish

Length: To 30 mm.
Distribution: Indo-Pacific, including all of our area.
Depth: 10-40 m.
General: Associates mainly with featherstars, living amongst tentacles but often swims around them. Occasionally with long-spined urchins. It is not as common as above species and differs with the extra line on the side and a pointed snout. Also seems to live deeper and with a preference for dark hosts.

Discotrema lineata (Briggs, 1966)

LONGTOMS • BELONIDAE

A family of surface fishes with about 10 genera and over 30 species. Their jaws are extremeky elongated with numerous needle-like teeth for which they are sometimes called neeedlefishes. Some species get large, 1.3 m and a weight of 5.2 kg, but a few which have adapted to freshwater are about 10 cm fully grown. They hunt small fishes in small packs or schools, but in turned are preyed on by the larger pelagics such as tuna and also dolfins. Most species are pelagic and venture far off shore. A few live in coastal zones where they patrol over shallow reefs, seagrass beds and along mangroves.

Longtoms are potentially dangerous when attracted by lights at night and there are known fatalities of leaping fish killing fisherman. Although no serious accident has been reported with divers, collisions at night have occurred.

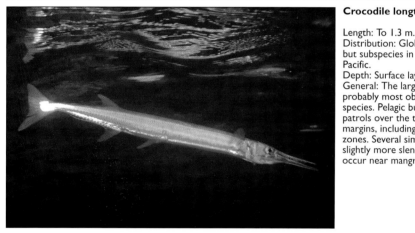

Crocodile longtom

Length: To 1.3 m.
Distribution: Global, tropical, but subspecies in eastern Pacific.
Depth: Surface layer.
General: The largest and probably most observed species. Pelagic but often patrols over the top of reef margins, including coastal zones. Several similar but slightly more slender species occur near mangroves.

Tylosurus crocodilus (Peron & Lesuer, 1821)

GARFISHES • HEMIRAMPHIDAE

A primarily tropical family with about 12 genera and 80 species. They are surface fishes typified by the elongated to greatly extended lower jaw in most species. They usually school in coastal and estuarine waters and some have adapted to freshwater.

Garfishes are mostly shiny and silvery, feeding on a variety of algae and zooplankton. Most species produce eggs with sticky filaments which quickly attach to floating matter, or sink when produced by those who spawn over seagrasses, however a few are thought to be viviparous. The two widespread and in our area commonly observed species are included here.

GARFISHES • HEMIRAMPHIDAE

Barred garfish

Length: To 35 cm.
Distribution: Indo-Pacific, from Africa, Red Sea to Samoa, including all of our area.
Depth: Surface layer.
General: Coastal bays and lagoons over shallow sand or reef flats, usually in the vicinity of seagrass beds, swimming in small loose groups. Juveniles are barred and large adults have a series of spots along back.

Hemiramphus far (Forsskal, 1775)

FLYING FISHES • EXOCOETIDAE

A large epipelagic family with about 7 genera and over 60 species. With greatly extended pectoral fins with large surface areas these fishes can glide great distances over the surface, often aided by the waves, to escape predators below. We see these fishes usually from a moving boat when the leap, often in small groups, away from the front or sides. The species vary mainly with different colour patterns on the wing-like fins when viewed from above. Although often abundant in our area with many species, as divers we rarely see this fishes. Only when diving at night on a rare occasion, and usually only when offshore.

Included photographs show a small juvenile taken at night agains the surface in the Philippines and and adult initiating its glide as it accelerate with the large lower lobe of the tail, shown by the pattern of the surface behind.

Cypselurus sp

FLASHLIGHTFISHES • ANOMALOPIDAE

A small interesting family with only two genera and five species which possesses a luminous organ immediately below the eye. This organ produces a very bright light, enough to able the fish to find and recognise prey when feeding on zooplankton at night. In addition it can be turned off or flashed on and off by turning it over or drawing a black skin across. This could serve to hide from an enemy or used to signal each other. During the day these fishes are well hidden in reefs and usually in deep water. At night they come out and rise to shallower depth where the food is, especially when very dark without moon or thick cloud cover.

The two species in our area are easily identified by the presence of one or two dorsal fins. Finding these fishes at night is by turning off the light or pointing to light away from the viewing area. This is how the juvenile below was found. Diving a few hours later rather than just after dark also improves chances to see them as they move to shallower depth with time. In some areas where they are known to occur, it is worthwile to make a special effort to see these fishes as they are a truly spectacular sight.

Two-fin flashlightfish

Length: To 28 cm.
Distribution: Indonesia, Malaysia, Philippines to southern Japan and Micronesia.
Depth: 20-400 m.
General: A very deep water species which occasionally is seen in groups in depths of about twenty metres in very long reaching caves or at night. It is easily recognised by the double dorsal fin, and rotate its organ to turn it on or off. The juvenile in the photo was found at 20 m depth on a very dark night, swimming solitair about 1 m above the substrate, flashing its light regularily with about one second intervals. This species is thought to comprise two forms, a shallow and a deep water form. In the shallows (about 20-50 m) it grows to only about 15 cm, but in the deep range it attains 28 cm. Examinations of specimens have failed to find any differences, other than size, and at this tage itis not known if there is a single species in which large adults live deep, or there are subspecies, or even different species.

Anomalops katoptron Bleeker, 1856

FLASHLIGHTFISHES • ANOMALOPIDAE

One-fin flashlightfish

Length: To 11 cm.
Distribution: Indonesia, Malaysia, Philippines to central Pacific.
Depth: 10-50+ m.
General: Coastal reefs with deep drop-offs, usually in large groups feeding just out from the wall in currents at night. The single dorsal fin readily identifies this species. It has a skin to cover the light-organ. Almost never seen during the day, except in the very dark parts of ship-wrecks or caves. Rises from deep water on dark moonless nights to shallow depths.
A small species which is easily maintained in aquaria, however it is extremely sensitive to changes of light levels or bright lights.

Photoblepharon palpebratus (Boddaert, 1781)

SQUIRRELFISHES • HOLOCENTRIDAE

A large, globally distributed tropical family with two distinct subfamilies: Holocentrinae with three genera, known as squirrelfishes, which has a prominent spine (thought to be venomous in some species) on the lower corner of the gill cover, and Myripristinae with five genera and known as soldierfishes, in which the gill-cover spine is small. In all there are about 70 species, mostly of red colouration and about 40 occur in our area. They are nocturnal but they are commonly seen during the day in caves or under large overhangs along drop-offs. Some occur solitair, are secretive, whilst others school in caves or near coral bommies. At night they hunt for shrimps or large zooplankton and swim individually over reefs or sand adjacent.

The smallest larvae known are about 2 mm long. The larvae of Holocentrinae have a long spiny rostrum at eye level which is lost or just evident in postlarvae. They settle when about 40-50 mm long but some pelagic specimens about 60 mm were photographed during a night dive off Sulawesi.

SQUIRRELFISHES • HOLOCENTRIDAE

Giant squirrelfish

Length: To 45 cm.
Distribution: Indo-Pacific, throughout our area.
Depth: 6-100+ m.
General: Shallow coastal to deep outer reefs. During the day in caves, often in pairs. The largest squirrelfish and known under many common names: 'spiny-, sabre-, and long-jawed squirrelfish', but as it is the largest species we prefer 'giant'.

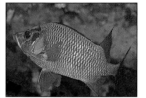

Sargocentron spiniferum (Forsskal, 1775)

Redcoat squirrelfish

Length: To 27 cm.
Distribution: Indo-Pacific, throughout our area.
Depth: 6-50 m.
General: Coastal reefs and large lagoons. Usually at moderate depths in small to large aggregations on gentle slopes and flats with large coral bommies. In some areas, especially remote oceanic islands, it is replaced by the similar but darker striped *S. praslin* (Lacépède, 1802).

Sargocentron rubrum (Forsskal, 1775)

SQUIRRELFISHES • HOLOCENTRIDAE

Finelined squirrelfish

Length: To 20 cm.
Distribution: Oceanic islands, Indian and Pacific, including north-eastern Indonesia, Philippines.
Depth: 3-50 m.
General: Rocky, rather than coral based reef in protected bays and lagoons but mainly restricted to oceanic margins in our area. Very secretive, well in the back of low ledges, hunting solitair at night. Best identified by two thicker white stripes sometimes in short streaks.

Sargocentron microstoma Günther, 1859)

Crown squirrelfish

Length: To 17 cm.
Distribution: Indo-Pacific, throughout our area.
Depth: 2-30 m.
General: Protected clear coastal and inner reefs along drop-offs and lagoons. Often seen during the day semi-exposed singly or in small aggregations with coral heads or in ledges.

Sargocentron diadema (Lacépède, 1802)

Samurai squirrelfish

Length: To 20 cm.
Distribution: Indo-Pacific, throughout our area.
Depth: 10-50 m.
General: Clear outer reef lagoons and slopes with remote outcops of rock and coral. Scattered localities in our area and seems to prefer deep water. Solitair, under ledges during the day.

Sargocentron ittodai (Jordan & Fowler, 1903)

SQUIRRELFISHES • HOLOCENTRIDAE

Sargocentron caudimaculatum (Rüppell, 1838)

White-tail squirrelfish

Length: To 25 cm.
Distribution: Indo-Pacific, throughout our area.
Depth: 6-50 m.
General: Coastal to outer protected reefs in lagoons and on walls in caves, singly or congregating into large groups during the day when they are easily recognised by the white tail, but at night turn bright red all over.

Sargocentron melanospilos (Bleeker, 1858)

Three-spot squirrelfish

Length: To 27 cm.
Distribution: Indo-Pacific, throughout our area.
Depth: 3-50 m.
General: Mostly oceanic zones in lagoons and clear coastal reefs which are semi-exposed to swell. Solitair in ledges and small caves in coral rich areas. Easily identified by the three black spots near the tail.

Sargocentron violaceum (Bleeker, 1853)

Violet squirrelfish

Length: To 25 cm.
Distribution: Indo-Pacific, throughout our area.
Depth: 6-30 m.
General: Clear coastal to protected inner reef slopes and walls with rich coral growth. Secretive during the day, usually just visible in small caves or in narrow ledges. Adults have a bright red head and is also known as red-face squirrelfish.

SQUIRRELFISHES • HOLOCENTRIDAE

Mouth-fin squirrelfish

Length: To 35 cm.
Distribution: Indo-Pacific, throughout our area.
Depth: 6-50 m.
General: Inner and outer reef walls in long wide ledges, singly or in small loose aggregations, usually in depths of 20 m or more. Has interesting dorsal fin pattern which looks like a large mouth with sharp white teeth which may serve for defensive bluff, flicking it upwards when approached.

Neoniphon opercularis (Valenciennes, 1831)

Blood-drop squirrelfish

Length: To 24 cm.
Distribution: Indo-Pacific, throughout our area.
Depth: 2-50 m.
General: Coastal to inner reef habitats, including sheltered lagoons, seagrass areas with coral patches and deep slopes with rich coral growth. Usually in groups with large branching corals, often hoovering just above. At night they scatter over the reefs to hunt on crustaceans.

Neoniphon sammara (Forsskal, 1775)

Silver squirrelfish

Length: To 24 cm.
Distribution: Indo-Pacific, throughout our area.
Depth: 15-50 m.
General: Coastal and protected inner reef slopes with thich coral growth in current prone areas. Usually with large coral heads, hoovering close in small groups. Looks very silvery in habitat.

Neoniphon argenteus (Valenciennes, 1831)

SOLDIERFISHES • HOLOCENTRIDAE

Myripristis murdjan (Forsskal, 1775)

Crimson soldierfish

Length: To 25 cm.
Distribution: Indo-Pacific, throughout our area.
Depth: 6-50 m.
General: A common but secretive species during the day, often confused with other similar of which there are several. Coastal, often silty rocky reefs and in shipwrecks. It differs from the next species by a red versus yellow spiny dorsal fin.
The other similar species *M. hexagona* is less secretive and often seen in schools in oceanic locations, and *M. amaena* lacks the white leading margins on the fins.

Myripristis berndti Jordan & Evermann, 1903

Yellow-fin soldierfish

Length: To 30 cm.
Distribution: Indo-Pacific, throughout our area.
Depth: 6-50 m.
General: One of the most widespread species, found on clear coastal and inner reefs along drop-offs in caves or with large spacious coral heads. Singly or in small groups, best recognised by yellow spiny dorsal fin.

Myripristis vittata Cuvier, 1831

White-tipped soldierfish

Length: To 20 cm.
Distribution: Indo-Pacific, throughout our area.
Depth: 10-50+ m.
General: Mostly found along deeper parts of outer reef drop-offs in caves, forming small to moderate sized groups.
Best recognised by white tipped spines in dorsal fin and the pale rather than dusky bar along gill-cover margin.

SOLDIERFISHES • HOLOCENTRIDAE

Violet soldierfish

Length: To 20 cm.
Distribution: Indo-Pacific, throughout our area.
Depth: 10-50 m.
General: A common species in our area along deep drop-offs, during the day in caves, forming small groups. Best recognised by the orange-tipped fins and silvery centred scales along their sides. Sometimes with black blotches on body and fins, once described as a separate species by Bleeker.

Myripristis violacea Bleeker, 1851

Splendid soldierfish

Length: To 25 cm.
Distribution: Indo-Pacific, throughout our area.
Depth: 15-50+ m.
General: Uncommon in our area and only occasionally seen in some areas, mainly oceanic or very deep. Clear deep lagoons and with large coral bommies, during the day hoovering just adjacent to their shelter, ready to disappear out of sight with danger. Singly or in pairs, however in some oceanic localities schooling.

Myripristis melanosticta Bleeker, 1863

Bronze soldierfish

Length: To 35 cm.
Distribution: Indo-Pacific, throughout our area.
Depth: 6-50+ m.
General: A distinct species with broad black margins on median fins. Looks very pale otherwise when viewed in natural habitat. Clear coastal and inner reefs, mostly in the deeper part of lagoons with coral bommies or in caves and ledges along drop-offs. In our area singly or in pairs, but elsewhere in oceanic locations reported to form schools.

Myripristis adusta Bleeker, 1853

SEAMOTHS • PEGASIDAE

A small Indo-Pacific family of small curious fishes, comprising 2 genera and 5 species. Small benthic fishes with horny depressed bodies and large horzontal kept pectoral fins, crawling on the bottom with the ventral and pectoral fins. A produced rostrum and inferior mouth protrusible downward and tube-like. Gill opening restricted to a small posterior openings facing upwards. Body encased in rigid bony plates and tail encircled by bony rings. Eggs and larvae are pelagic and small juveniles may float into tidal pools. Their diet consists primarily of tiny benthic crustaceans and other small invertebrates.

In some species there is sexual dimorphism in shape and colour of the pectoral fins. Males display intensified colours bordering these fins during the occasion by circling the female. Like pipefishes, the female rises with the male closely together towards the surface where they release eggs and sperm. Interestingly one can see how the evolution shaped the future of the pipefishes here, as it wouldn't take much for a few eggs to stick to the male, the basis for parental care in the pipefish family.

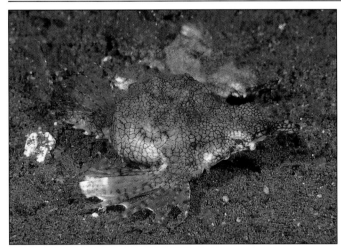

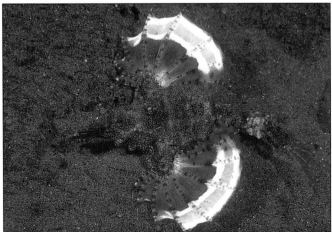

Little dragonfish

Length: To 8 cm.
Distribution: Indo-Pacific, including all of our area.
Depth: 3-15 m.
General: Colour highly variable, matching surroundings and capable of quick changes. Mimics bits of surrounding debris such as pieces of broken shell. Sheltered coastal bays on fine sand with sparse rubble. Extremely well camouflages and partly buries when resting. Regularly discards outer layer of skin in one complete piece, apparently to prevent accumulation of epibiotic growths such as algae, hydroids and other organisms.
Although moderately common in some areas, this species is rarely seen due to ist cryptic habits. Finding specimens is usually by accident but where there is one, there are usually more, especially when adult. They form loose pairs which are only found close when the time for spawning is near. Shown from top to below: Pair, male and small juvenile.

Eurypegasus draconis (Linnaeus, 1766)

TRUMPETFISHES • AULOSTOMIDAE

This family comprises only 1 genus and 2 species, divided between Atlantic and Indo-Pacific seas. Body very elongate and slightly compressed. Head compressed, snout tubular and produced. Dorsal fin with evenly spaced short spines, each separate with a small triangular membrane following. Soft rayed part of dorsal fin similar and opposite to anal fin, posteriorly placed, well back on the body near caudal fin. Ventral fins well back, about halfway along body. Caudal fin small and lanceolate. Cunning predators, positioning themselfs vertically along objects or rides immediately on the back of other fishes to be able to get close to prey, consisting primarily of other fishes. The Indo-Pacific species is wide-ranging in tropical seas and expatriates to warm-temperate zones.

Easily identified by their robust body and colour. The only similar species are the cornetfishes, FISTULARIDAE, which are very plain and more slender with a long filament on the centre of the caudal fin. The bright yellow phase or form is usually noticed most, although the normal brownish or grey form is more common. The yellow form seems to be more daring as well and juveniles are especially secretive, which apart from seagrass areas are often deep, found amongst large black coral fans.

Trumpetfish

Length: To 60 cm, doubtfully reported to 90 cm.
Distribution: Indo-Pacific, including all of our area.
Depth: 1-35 m.
General: Colour variable between juvenile and adult, and individuals, from longitudinal bands to vertical banding, and can quickly change colour. A xanthic form is common in some areas. Juveniles in seagrass and soft coral areas. Adults from shallow reef flats to deep drop-offs, and from coastal waters to outer reefs, mostly solitair, sometimes loosely in pairs. Colour variations shown. A predator feeding on other fish and the mouth can open surprisingly large, creating a strong suction in the process. Not a good candidate for the home aquarium, but no doubt this species, like other predators, plays an important role in bringing the best out of the species it prays on.

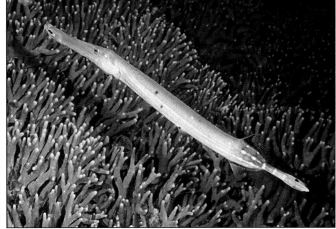

Aulostomus chinensis (Linnaeus, 1766)

FLUTEMOUTHS • FISTULARIIDAE

The flutemouths are a single genus family comprising 4 species, distributed globally in tropical and warm-temperate seas. Similar fishes with a long tubular body and long, slightly compressed head and extremely long snout with a relatively small mouth at the end. The middle pair of rays in the caudal fin is greatly produced as an extremely long filament, the connecting membrane large in juveniles. Excluding central filament, caudal fin forked. Ventral fin small well back from head, just ahead of anus. Fin spines absent. Coastal to off-shore reefs to deep continental shelf depths. Young enter estuaries. Swims close to the bottom, singly of in small aggregations, hunting small fishes as main food source. Large fishes, some reaching 2 m in length.

Although two species are known from our area, only one is commonly observed to adult size, at least during the day. The other species is common in shallow depths in the more temperate region of ist range but enters shallows in tropical waters as juvenile and occasionally at night to hunt.

Smooth flutemouth

Length: To 1.5 m.
Distribution: Indo-Pacific, throughout our area.
Depth: 1-100+ m.
General: Adults with thin blue line or dashes dorsally. Coastal reef flats and lagoons, often in small groups.
Second species, the rough flutemouth maybe encountered in divable depths and only differs at small size in having a round pupil (shown with diver below). At night both species show a broadly banded pattern which can be turned off quickly it kept in the light. The smooth flutemouth occurs in small groups which are often found in open parts of reef crests, however these are usually sub-adults. The large and fully grown adults prefer deep water and are only seen during the night in waters adjacent to deep water from which they rise to hunt prey in shallow depths.

Fistularia commersonii Rüppell, 1838

SHRIMPFISHES • CENTRISCIDAE

A small family of perculiar fishes which have adapted to a vertical position in life, swimming with head down. Two genera and four species are recognised in the Indo-Pacific. They can be found in pairs or large schools comprising countless individuals, swimming in a synchronised fashion, all turning at the same time. Although in general only one species is thought to be common on reefs in our area, there are two similar species which are easily separated because of the differences in their large spine in the dorsal fin, rigid in one and hinged in the other.

Both are commonly encountered in schools in our area but because of identification problems, their true distribution may not be widespread, and in some areas only one species maybe present. Juveniles of both species may seek shelter amongst spines of long-spined sea-urchins, adults occur in pairs or form schools in soft-coral zones, especially where seawhips are in dense patches.

They feed on tiny crustaceans which swim near the substrate such as mysids (possum shrimps, some full size at only 5 mm long). Eggs are probably pelagic like their off spring which have a short snout, developing into a more adult like shape at about 10 mm when they are found in surface waters, fully pigmented, swimming amongst loose weed, ready to settle on the substrate when they drift to shore, usually amongst urchin spines or other spiny echinoderms.

Both species adapt easily to aquaria but like pipefishes and relatives their food is small crustaceans and as long as the foodsource is no problem, they will thrive.

Coral shrimpfish

Length: To 14 cm.
Distribution: Indo-Pacific, containing all of our area.
Depth: 1-30 m.
General: Clear coastal reefs and lagoons, schooling over branching corals and often in lagoons in small schools above grouping long-spined urchins. Replaced in more silty area by the rigid shrimpfish, **C. scutatus,** on the next page. Tiny juveniles, only about 5 mm long, may settle between the protective spines of echinoderms, including *Diadema* urchins or crinoids, and may not be readily recognised. Post larvae have a proportionally much shorter snout than the adults, though this quickly grows after settling. They seek each others company and form small schools as they find each other.

Aeoliscus strigatus (Günther, 1860)

SHRIMPFISHES • CENTRISCIDAE

Rigid shrimpfish

Length: To 15 cm.
Distribution: Indo-Pacific, throughout our area.
Depth: 3-100+ m.
General: Coastal, often muddy habitat, to deep off shore. Singly, pairs or in large dense schools. Individuals often found sheltering next to seapens or strapweed out in the open. Schools in rich coral patches with fans or in seawhip gardens.

Centriscus scutatus Linnaeus, 1758

GHOST PIPEFISHES • SOLENOSTOMIDAE

A small tropical family closely related to Syngnathidae, the true pipefishes, comprising a single genus with 5 species. They have short compressed bodies which are encased with bony plates, two separate dorsal fins, large anal fin, similar and opposite to second dorsal fin, and a large ventral fin which in the female is enlarged and hooked onto the body to form a pouch for holding eggs. The eggs are small, less then 1 mm in diameter and numerous, stuck to the inside of each fin. After about 3 weeks the tiny young hatch singly or a few at the time, usually during the fanning which is about as regular as the rhythm of the gills. It appears that they have a prolonged larval stage and settling juveniles are over half adult size and ready to breed. Young adults are more slender than older specimens which in growth deepen at the caudal peduncle, the body between dorsal and ventral fins and in the snout. Such changes has led to the various forms being described as different species. Ghostpipefishes are only found in quiet environments, usually in coastal bays, hoovering vertically with head down, near soft corals, fans on walls or seagrasses. They feed on small shrims and mysids.

Robust ghostpipefish

Length: To 15 cm.
Distribution: Indo-Pacific, throughout our area.
Depth: 3-25 m.
General: Colour variable from green to brownish-red, or blackish. Common in seagrass bed areas but very well camouflaged. Also on deeper coastal reefs in 15 to 25 m and usually are reddish brown or blackish there.

Right:
Solenostomus paradoxus pairing in black coral, Bali, Indonesia.

Solenostomus cyanopterus Bleeker, 1854

GHOST PIPEFISHES • SOLENOSTOMIDAE

Ornate ghostpipefish

Length: To 10 cm.
Distribution: Indo-Pacific, throughout our area.
Depth: 3-25 m.
General: Variable from red, white or yellow blotching per segment, to almost totally black (see next page). Coastal reefs with rock faces or coral drop-offs, near the bottom, floating vertically. A common species, but easily overlooked.

A rarely observed occasion of the birth of the ghost pipefish. The tiny hatchling just expelled from the pouch with other eggs ready to hatch visible. Photograph taken off Flores, Lesser Sunda islands, Indonesia, in only 3 m depth.

Solenostomus paradoxus (Pallas, 1770)

PIPEFISHES & SEAHORSES • SYNGNATHIDAE

The family Syngnathidae comprises some of the most interesting fishes in both behaviour and appearance. This large family comprises over 50 genera and well over 200 species, distributed globally in all but the coldest seas and some occur in freshwater. There are two subfamilies, based on absence or presence of a caudal fin, including larval stages. Most pipefishes belong in Syngnathinae, but some and the seahorses belong in Hippocampinae. Tropical Seahorses are badly in need of revision with many similar species.

Mostly slender fishes, bodies encased in bony plates, arranged in series of rings. A small oblique mouth at the end of a tubular snout. Gills are enclosed with a small pore-like opening above the opercle. Ventral fins and jaw teeth are absent. Anal and caudal fin may be very small or lacking in certain species or depending of its stage. Tail is prehensile in some species. Usually the species live on shallow reefs or in seagrass beds, feeding primarily on small crustaceans which are sucked up with the long snout. They range through a broad range of depths, in protected bays some species occur as shallow as the intertidal zone, and off-shore some species are trawled in excess of 400 m. Sizes also have a broad range with some species maturing at about 25 mm and the largest species exceeds 65 cm. Their reproduction methode is unique, as males incubate the eggs in a pouch or ventrally on the tail or belly, with or without skin cover. Hatchlings are often well advanced and may already resemble its parents.

Seahorses and pipehorses are used as souvenirs and by Chinese people for medicinal purposes, believing that it prevents aging of the skin. Most likely the opposite is true(!) Millions of *Hippocampus kuda* are taken from Philippine and Indonesian waters every year and it has greatly effected the seahorse populations in many areas, particularly in the Philippines where fully grown specimens are now almost non-existing. This senseless activity is difficult to stop as long there is a demand. Changing centuries old habits can not be done overnight and it seems the best way to protect the wild population is to set up fish farms to breed seahorses in great numbers, but until now there have been few attemps which failed.

Common seahorse

Length: To 28 cm.
Distribution: Indo-Pacific, all of our area but several similar species involved ('kuda' complex).
Depth: 0.5-30 (?) m.
General: Juveniles with tiny black spots peppered all over, adults with low crown on top of head and with dark speckles and pale scribbles, but varies between different habitats. Sheltered coastal bays and lagoons in seagrass beds, usually on the edges clinging on to sticks or something else ridged. Also attached to floating weeds on surface.

Hippocampus kuda Bleeker, 1852

SEAHORSES & PIPEHORSES • SYNGNATHIDAE

Hippocampus hixtrix Kaup, 1856

Thorny seahorse

Length: To 15 cm.
Distribution: Indo-Pacific, throughout our area.
Depth: 1-20 m.
General: Seagrass beds and weedy low rocky reef in the shallows and in sponge areas deeper.
Lives in pairs, staying in close vicinity of each other.

Acentronura tentaculata Günther, 1870

Dwarf pipehorse

Length: To 63 mm.
Distribution: Indo-Pacific, throughout our area.
Depth: 1-10 m.
General: As photograph shows, incredible camouflage, and combined with size it will take a trained or eager eye to spot one of these. Still coastal sand flats, attached to algae and small plants. The male has a pouch similar to a seahorse.

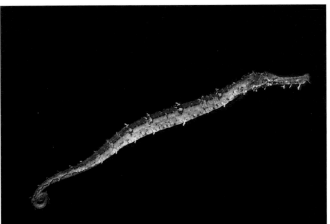

Syngnathoides biaculeatus (Bloch, 1785)

Double-ended pipehorse

Length: To 30 cm.
Distribution: Indo-Pacific, throughout our area.
Depth: 1-10 m.
General: Weedy habitats in protected coastal bays and lagoons, with broad-leaf seagrasses and in floating weeds, including sargassum rafts in open sea. Male carries eggs openly on trunk. Closely related to *Solegnathus*, the deep water pipehorses and seadragons from Australia.

PIPEFISHES • SYNGNATIDAE

Long-snout pipefish

L: To 32 cm. Di: Indo-Pacific, including all of our area. De: 15-90 m. Ge: In clear sandy bays, lying low on substrate near algae or bits of loose weed. A second similar species *T. bicoarctatus* with head more at angle.

Trachyrhamphus longirostris Kaup, 1856

Schultz's pipefish

Length: To 15 cm.
Distribution: Indo-Pacific, Red Sea to Samoa, including all of our area.
Depth: 1-30 m.
General: A common species on reef crests, sandy lagoons with coral patches and rubble flats. Often on living corals and sometimes in small aggregations on sand when in pursuit of prey.

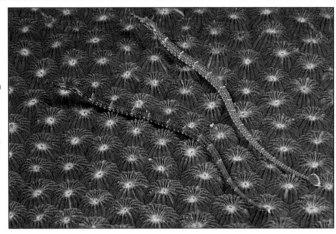

Corythoichthys schultzi Herald, 1953

Whiskered pipefish

L: To 18 cm. Di: Widespread tropical Indo-Pacific. De: 1-25 m. Ge: Sheltered sand slopes with sparse algae or plant growth, usually in depths of about 6-10 m. Lies low on the substrate and extremely well camouflaged. Juveniles have a series of large leafy appendages on back which reduce with growth.

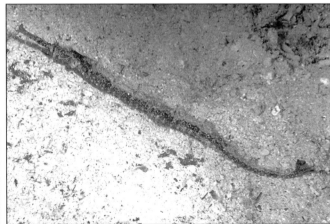

Halicampus macrorhynchus Bamber, 1915

PIPEFISHES • SYNGNATIDAE

Banded pipefish

Length: To 20 cm.
Distribution: West Pacific, probably all of our area but maybe replaced by *D. multiannulatus* from Sumatra west.
Depth: 1-50 m.
General: Protected coastal reefs and lagoons, in caves or amongst boulders, especially when long-spined urchins present. Singly, pairs and sometimes forming large aggregations along bases of boulder reefs with urchins. Newly laid eggs are red and deposited to the underside of male without additional cover, typical for free-swimming species.

Doryrhamphus dactyliophorus (Bleeker, 1853)

Multibar pipefish

Length: To 18 cm.
Distribution: Indian Ocean, east to Sumatra, Indonesia.
Depth: 3-45 m.
General: Like above species in caves and crevices, usually living in pairs. Photograph in Andaman Sea which also show a pair of **Blue-stripe pipefish** *D. exicus* a smaller species often with *Diadema* urchins.

Doryrhamphus multiannulatus (Regan, 1903)

PIPEFISHES • SYNGNATIDAE

Jans's pipefish

Length: To 13 cm.
Distribution: West Pacific, ranging to Andaman Sea, and found in all of our area.
Depth: 3-35 m.
General: Usually a secretive pipefish in rich coastal reefs under large plates or in dense bushy sponges, however an active cleaner in some areas where pairs swim upside down under large table corals and specialise in cleaning cardinals, *Cheilodipterus spp*, and damsels, *Neopomacentrus spp.* Often seen in the background of caves where cleaner shrimp work, and no doubt the pipefish gets involved when the right customers arrive for treatment.

Doryrhamphus janssi (Herald & Randall, 1972)

Mushroom-coral pipefish

Length: To 80 mm.
Distribution: Only known from Indonesia and Philippines.
Depth: 10-20 m.
General: Lives exclusively in association in *Heliofungia* mushroom corals which expend their tentacles during the day' looking like an anemone. Entire families of this pipefish can be found together on a single coral. Clear inner reefs, along bases of shallow drop-offs.

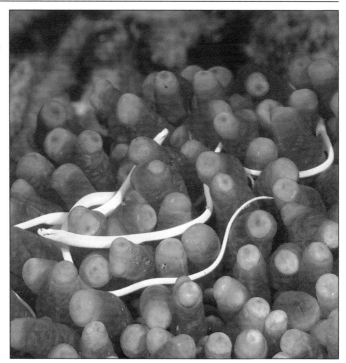

Siokunichthys nigrolineatus Dawson, 1983

SCORPIONFISHES • SCORPAENIDAE

A complicated family with about 10 sub-families, 2 of which comprise our main species: PTEROINAE, the lionfishes, and SCORPAENINAE, the scorpionfishes or rockcods. The lionfishes are the most spectacular with very long dorsal spines and great pectoral fins, often reaching past anal fin. The scorpionfishes are a diverse group, comprising at least 12 genera which are widely distributed in all but the coldest seas. They have spiny heads and are mostly substrate hugging and very well camouflaged fishes. Dispite their appearance most species are excellent food fish and in many areas some species are of commercial value. The lionfishes are often kept in aquaria because of their interesting appearance, but one has to be careful in choosing companions as they have a large mouth and small fish will quickly disappear.

The lionfishes are distinct in appearance and most species are easily identified, however even the common lionfish in the Indian Ocean was confused with another species (here rectified). The scorpionfishes are highly camouflaged and the species within a genus are almost identical and sometimes a specimen's body is needed to make a positive identification possible. A great number of scorpionfishes are small, less than 10 cm long, and very secretive in ledges or under corals but some come out at night.

WARNING: All species possess venomous spines and a sting produces extreme pain, followed by numbness, and although rare the stonefish (sub-family SYNANCEIINAE) has caused fatalities. If stung, the best thing to do is to apply heat to the wound as soon as possible. Immersing the wound in very hot water is generally recommended but hot air from a hair-dryer or heater-radiator maybe more practicle, can be just as effective, and can be controlled better. Heat kills the venom, when applied to the wound and near surroundings, usually the relied of pain is almost instantly.

Eggs and larvae are pelagic, generally very small, eggs about 1 mm round or slightly ovoid eggs, hatching larvae about 2 mm long. Postlarvae can be less than 10 mm long depending on the species, e.g *Pterois* as small as 15 mm TL has been found on the substrate. All the scorpaenid species are carnivores, some praying on other fishes only, but others may prefer crustaceans or a variety of both. Whilst many species are very interesting or spectacular and easily kept in an aquarium, it should be kept in mind that they generally have a large mouth and have a taste for expensive species.

Spotfin lionfish

Length: To 20 cm.
Distribution: Indo-Pacific, common throughout our area.
Depth: 6-50 m.
General: Two forms which may represent different species. One form lives deep and away from reefs, often muddy substrate, the other form on reefs which is usually encountered by divers and included here. Banding usually pale to dark brown with variable width and pectoral fin rays white. Coastal to offshore reefs from shallow reef flats to deep slopes in coral heads. Feeds on fishes and crustaceans.

Pterois antennata (Bloch, 1787)

SCORPIONFISHES • SCORPAENIDAE

Pterois radiata Cuvier, 1829

White-lined lionfish

Length: To 25 cm.
Distribution: Indo-Pacific, including all of our area.
Depth: 1-15 m.
General: A distinct species with the dark body colour and white lines. Unusual in its preference for habitat which is rock reef with limited coral growth, seemingly sensitive to their stings, and therefore rarely seen in coral rich areas. Hides in ledges and caves during the day, coming out on dusk to feed primarily on crustaceans.

Dendrochirus brachypterus (Cuvier, 1829)

Dwarf lionfish

L: To 15 cm. Di: Indo-Pacific, throughout our area.
De: 10-100 m. Ge: Extremely variable in general colour from red to brown or dark purplish-brown, and a rare yellow form. Coastal reefs and estuaries, often silty environment.

Dendrochirus zebra (Cuvier, 1829)

Zebra lionfish

L: To 20 cm. Di: Indo-Pacific, throughout our area.
De: 3-80 m. Ge: Coastal reefs and lagoons, in large coral heads, shallow to fairly deep water. *Dendrochirus* differs only slightly from *Pterois* in shorter rays and divided pectoral fin rays in adults.

SCORPIONFISHES • SCORPAENIDAE

Common lionfish

L: To 35 cm. Di: Tropical Indo-Pacific, from east Africa to west Pacific, throughout Southeast Asia. De: 1-50 m. Ge: A common species on reefs, often seen hovering in caves or near coral heads. Small juveniles are more secretive in ledges and occur singly whilst adults form small groups. Shows some variation in colour, typically zebra striped (right below, Banda Sea) in general but sometimes entire fish very dark (right, Andaman Sea). Such dark specimens placed in an aquarium changed to the lighter colour in a short time. Recently the Indian Ocean population was thought to be distinct in the basis of colour and fin ray count, however this is wrong. P. volitans is recognised by the spotting of the median fins, which is very distinct. These fishes are also known as Firefishes, Devilfishes and Turkeyfishes, names in relation to their large feathery pectoral fins or venomous and painful stinging dorsalfin spines. They feed primarily on other fishes, chasing them even in groups.

Pterois volitans (Linnaeus, 1758)

Soldier lionfish

Length: To 35 cm.
Distribution: Tropical Indo-Pacific, from east Africa to west Pacific, throughout our area. Depth: 10-100 m.
General: Because of it´s similarity to the Common lionfish, this species is often not noticed. It prefers deeper water and mainly coastal, often silty, habitats. Lionfish on deep remote outcrops in still coastal mud or sand flats are usual this species. It is more brownish than the Common lionfish and shows little or no spottings on the median fins.
P. russelli is a synonym of P. miles!

Pterois miles (Bennett, 1828)

SCORPIONFISHES • SCORPAENIDAE

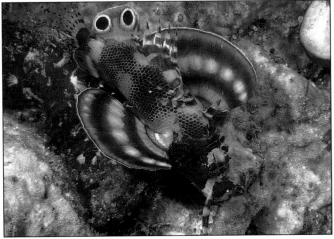

Two-eyed lionfish

Length: To 20 cm.
Distribution: Indian Ocean from Andaman Sea to West Pacific, throughout our area.
Depth: 3-50 m.
General: A very distinct species with the ocelli in the soft dorsal fin and the long barbels on the snout. Rarely seen during the day, but commonly seen at night in many area on shallow coastal reef flats and in caves, thus mainly nocturnal.
Usually seen singly but photograph show a pair during courtship on dusk in which the male can be recognised by the grey ocelli.

Dendrochirus biocellatus (Fowler, 1938)

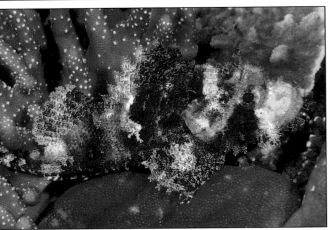

Raggy scorpionfish

L: To 25 cm. Di: Indo-Pacific, throughout our area.
De: 3-55 m. Ge: Several very similar species, all well camouflaged to suit surroundings. Mainly coastal reefs with soft corals and sponges.
On coral reefs, **S. oxycephalus** (below).

Scorpaenopsis venosa (Cuvier, 1829)

SCORPIONFISHES • SCORPAENIDAE

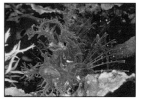

Lacy scorpionfish
L: To 25 cm. Di: New Caledonia through eastern New Guinea, Philippines to southern Japan. De: 6-25 m. Ge: A rare but remarkable species with a distinct appearance. Extremely well camouflaged dispite sitting out in the open on corals or sponges, seemingly disguised as a crinoid. Variation below.

Rhinopias aphanes Eschmeyer, 1973

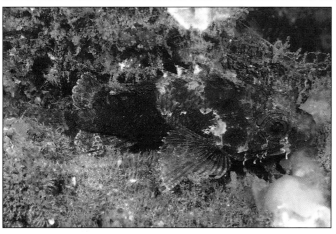

Blotchfin scorpionfish

Length: To 12 cm.
Distribution: Indo-Pacific, throughout our area.
Depth: 3-20 m.
General: A secretive species, in caves and ledges. Usually out at night.
Distinct from several other similar species ba the red markings on the fins and head.

Scorpaenodes varipinnis Smith, 1829

Shortfin scorpionfish

L: To 85 mm. Di: Indo-Pacific, throughout our area.
De: 10-50 m. Ge: A small species, best recognised by the large pale area over the body. Usually only out at night and often on the side of sponges.

Previous page:
The small white spots on the face of the lacy scorpionfish detract from the eyes which are like the rest of the body part of camouflage.
It may give the impression that this predator has poor sight and is harmless.

Scorpaenodes parvipinnis (Garrett, 1864)

SCORPIONFISHES • SCORPAENIDAE

Paper fish

Length: To 12 cm.
Distribution: Indo-Pacific, throughout our area.
Depth: 10-100+ m.
General: Extremely variable in colour, matching sponges and algae from white, green, brown, red, and various spotting. Settling juveniles at about 20 mm long are semi-transparent and have a smooth skin. A common species in many areas but is so well camouflaged that they are easily missed, eventhough specimens sitting on corals can be obvious. On coral reefs usually on shallow reef crests and slopes with sparse vegetation or sponges and has been recorded from trawls to 135 m. Variations shown.

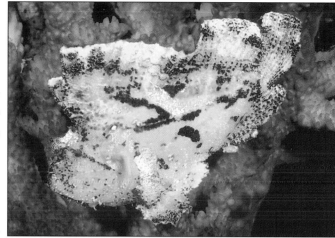

Taenianotus triacanthus Lacépède, 1802

Reef stonefish

Length: To 38 cm.
Distribution: Indo-Pacific, throughout our area.
Depth: 3-50 m.
General: The most venomous reef fish, producing incredible pain instantly and has caused fatalities. The strong fin spines have large amounts of venom at their base and when a spine penetrates (even foot wear) the skin is pushed down around the spine which causes pressure on the venom which then shoots upwards into the spiked area.
See family text for treatment.

Synanceia verrucosa Bloch & Schneider, 1801

WASPFISHES • TETRAROGIDAE

A moderate sized family of small and little known species, usually included with the SCORPAENIDAE as an easy way out. It comprises an estimated 15 genera and about 40 species. They feature a large dorsal fin which originates on top of the head and is often confluent with the caudal fin. Only a few species are encountered by divers because of their secretive nature and usually seen at night. Like the scorpionfishes they can produce painful stings, though the amount of venom would probably be small in comparison because of smaller size in general.

Only one genus is included here as they are the main ones seen and can be found in the open during the day as well. Three similar species occur in our area with an Indian and Pacicific species and the third with a restricted range in Indonesia. They are well camouflaged with their leaf-like disguise and when disturbed rock slowly sideways as pushed by a surge. It also does this to confuse prey which are mainly shrimps and other small invertebrates. Because of their general leaf-like appearance and crested dorsal fin they are known as leaf fishes.

Spiny leaf fish
L: To 18 cm. Di: Indonesia, but probably ranging to Malaysia and Philippines. De: 10-50 m. Ge: Coastal sand and mud slopes adjacent to deep water. This species is very similar to *A. binotatus,* Indian Ocean (below), with the highly crested dorsal fin but has many more rays in the anal fin.

Ablabys macracanthus (Bleeker, 1852)

Cockatoo leaf fish

Length: To 15 cm.
Distribution: Tropical west Pacific, Indonesia, Philippines, Malaysia, ranging to warm-temperate zones of Australia and Japan.
Depth: 3-20 m.
General: Protected coastal bays on sand and rubble-algae reef, usually seen on dusk out in the open on shallow sand flats with sparse seagrasses. Indian Ocean records are misidentifications and photographs taken in the west Pacific are often used in Indian Ocean fish books, adding to the confusion.

Ablabys taenianotus (Cuvier, 1829)

FLYING GURNARDS • DACTYLOPTERIDAE

A small scorpaenoid family comprising only two genera and seven species, one of which commonly seen in our area and thus included here. Their most obvious feature is the great pectoral fins which reach to the end of the caudal fin, and when spread form an almost round disc. They use the large pectoral fins for display, sexual, territorial and also to startle possible predators. The fins are folded along the side when swimming fast, but spread for further gliding along, usually just above the substrate in a flying fashoin. They do not fly through the air like the flying fish.

Small juveniles make interesting aquarium pets but it should be kept in mind that they have a large mouth and they hunt small fishes such as gobies. They also prey on a large variety of invertebrates which are probed for with their fingerlike separated pectoral-fin rays from the sand. Eggs are pelagic and the smallest juveniles known about 25 mm long.

Flying gurnard

Length: To 32 cm.
Distribution: Very widespread, including all of the tropical Indo-Pacific from east Africa to the west Pacific, throughout our area.
Depth: 6-100+ m.
General: A spectacular species when displaying. Well camouflaged when resting and may partly bury in sand.
Protected coastal bays on shallow sand flats and slopes with sparse reef or plant life, occasionally in pairs. Known also from trawls in excess of 100 m depth.

Dactyloptena orientalis (Cuvier, 1829)

FLATHEADS • PLATYCEPHALIDAE

A large, mainly Indo-Pacific family with numerous small species in the tropics and medium sized ones in warm-temperate zones where they are of great commercial importance. There are about 18 genera and 60 species, though only a few are regularily observed by diving. Nearly all species bury in sand and/or are extremely well camouflaged, many of which restricted to moderate depths or muddy estuaries. They are spiny scorpaenoid fishes with greatly depressed heads, armed with serrated ridges and pungent spines, usually enlarged on the sides, though not known to carry venom.

The flatheads with few exceptions lack swimbladders and spent nearly all their time resting on the substrate. They are carnivores which ambush prey from their buried position with just the eyes exposed, or feed at night by crawling and swimming along the bottom. The diet varies between species, some preferring molluscs, others crustaceans or fish, and some are not fussy at all, taking virtually anything which comes in striking range.

Some species migrate and congregate annually to breed, often seen in pairs during that time. Eggs are pelagic, round and small, less than 1 mm in diameter.

Fringe-lip flathead

Length: To 25 cm.
Distribution: Widespread tropical Indo-Pacific, including all of our area.
Depth: 1-30 m.
General: A common species in protected shallow sandy bays. Buries during the day and hunts at night when they can be seen sitting out in the open on the sand.
There are several more similar species, some with a longer snout and others differ in the spiny ridges on the head. It takes an expert to tell them apart and included here one of the most encountered species.

Thysanophrys otaisensis (Parkinson, 1829)

Crocodile fish

Length: To 50 cm.
Distribution: Singapore, Malaysia and Indonesia, ranging north to southern Japan and Micronesia.
Depth: 1-30 m.
General: A large but well camouflaged species which often sits on top of reef flats. Mainly protected coastal bays from mangroves to reefs and in lagoons.

Opposit page:
Cephalopholis sonnerati visiting a *Lysmata amboinensis* cleaning station.

Cymbacephalus beauforti (Knapp, 1973)

ROCKCODS, GROUPERS AND BASSLETS • SERRANIDAE

The Serranidae are a highly diverse family with several major groups, but classification is continuously under review. Changes can be expected with regards to family status. As presently defined, there are nearly 50 genera and well over 400 species, distributed world wide. They are here presented in various sub-families: Epinephelinae, the groupers and cods; Anthiinae, the basslets; Grammistinae, the soapfishes. Even within these groups, especially the latter, they are diverse assemblages of fishes, varying in size and shapes, or behaviour.

The Epinephelinae are the large rockcods (including the largest reef-fish of them all: the Queensland groper), gropers or coral-trouts. Most of the larger species are of commercial importance, reaching 1-2 m. They feature tiny, often embedded scales and as adults have small egg-shaped eyes. Being home-ranging on reefs, some are vunerable to spearfishing, causing decline of species in certain areas. Most live solitair and are territorial, claiming caves and the larger species claiming the greatest reef sections. Some species may congregate in certain areas, at least annually. Diet comprises a variety of fish, cephalopods and crustaceans.

The Anthiinae are the basslets or seaperches, the small tropical species, most of which schooling planktivores but a few live on the substrate similar to the rockcods. They feature small to moderate sized ctenoid scales and large eyes, placed laterally and anteriorly on head. The caudal fin is usually deeply emarginate to greatly lunate. In the planktivores, the males are often brightly coloured, and entirely different from females. Females greatly outnumber males and males derive from the larger usually most dominant females which take charge of groups of females in a harem fashion.

The Grammistinae or soapfishes have a slimy skin which contains a toxic. It comprises the most diverse group within the Serranidae and several genera will no doubt eventually give rise to a subfamily or even family status, especially the reef-basslets. Presently these different assemblages are referred to as tribes. The soapfishes are usually more secretive than other serranids, living in caves or ledges, remaining close or hidden in the substrate. Some species can be found in loose groups and show little fear towards divers.

Tomato cod
L: To 65 cm. Di: Indo-Pacific. De: 10-150+ m. Ge: Very small juveniles black with blue head and broad white margin on caudal fin, changing to brown and increasing orange spotting, to bright red adults. Coastal slopes with rocky or coral outcrops on rubble-mud habitat. Adults usually deeper.

Cephalopholis sonnerati (Valenciennes, 1828)

ROCKCODS • SERRANIDAE

Orange rockcod

Length: To 30 cm.
Distribution: Indo-Pacific, throughout our range.
Depth: 10-100+ m.
General: Mainly seen along deep drop-offs on outer reefs, in caves and along ledges. Usually fairly deep, 30 m or more, and with natural light looks a pale colour with the red filtered out. Similar to *C. aurantia* a deep water species which grows much larger.

Cephalopholis spiloparaea (Valenciennes, 1828)

Coral rockcod

L: To 40 cm. Di: Indo-Pacific, throughout our range.
De: 3-50+ m. Ge: Clear coastal to outer reefs. Often in small groups. Mainly rich coral growth along slopes and drop-offs, often in shipwrecks. Small juveniles uniformly orange, secretive amongst corals.

Cephalopholis miniata (Forsskal, 1775)

Saddled rockcod

Length: To 45 cm.
Distribution: Indo-Pacific, throughout our area.
Depth: 10-150+ m.
General: Clear coastal to outer reef drop-offs, usually in large caves with sponges. Commonly seen upside down or positioned vertically on wall. Scientists checking gut contens found that they eat mostly apogons, anthiids and shrimps.

Cephalopholis sexmaculata (Rüppell, 1828)

ROCKCODS • SERRANIDAE

Cephalopholis urodeta (Forster, 1801)

Flagtail rockcod

Length: To 30 cm.
Distribution: Indo-Pacific, two forms: Indian and Pacific.
Depth: 1-40 m.
General: This species commonly occurs on coastal reef crests with rich coral growth, mixed soft and stony corals. Highly variable in colour, shown Pacific form. Indian Ocean form lacks the white stripes in the tail and occurs from Java west.

Cephalopholis leoparda (Lacépède, 1801)

Leopard rockcod

Length: To 25 cm.
Distribution: Indo-Pacific from broadly our area to east Africa and central Pacific.
Depth: 3-40 m.
General: One of the smallest rockcods and larger sizes reported are based on misidentifications. A secretive species but common in rich coral reefs from the crests to densely covered slopes. Best identified by the dark tail saddle and stripes in fin.

Cephalopholis argus Bloch & Schneider, 1801

Peacock rockcod

Length: To 45 cm.
Distribution: Indo-Pacific, throughout our area.
Depth: 1-45 m.
General: Clear coastal reef crests to outer reef lagoons. Among rocks and corals, often seeing swimming about during the day. Variable in colour and the white conspicious area around the pectoral fin base can be turned on and off like from a switch.

ROCKCODS • SERRANIDAE

Bluespotted rockcod
L: To 35 cm. Di: Ranging throughout our area, including to Andaman Sea. De: 1-50 m. Ge: Shallow coastal reefs to deep outerreef drop-offs. Caves and ledges, usually staying close to the substrate, moving through corals between areas. Juv. with bright yellow fins were once thought to be a different species.

Cephalopholis cyanostigma (Valenciennes, 1828)

Bluelined rockcod

Length: To 34 cm.
Distribution: West Pacific, throughout our area, including Andaman Sea.
Depth: 2-20 m.
General: This grouper is a shallow-water species of sheltered silty reefs, occuring also in the vicinity of river mouths.
Photo locality: submerged reef at the Indian Ocean coast off Thailand.

Cephalopholis formosa (Shaw, 1804)

Harlequin rockcod

Length: To 35 cm.
Distribution: Indo-Pacific, throughout our area, but mainly equatorial.
Depth: 10-70+ m.
General: At moderate depths, 30 m+, along clear water outer reef dropoffs with large caves and ledges, seemingly preferring good sponge growth. Stayes in caves and below overhangs but can be approach at close range. Juveniles are pink with fewer lines.

Cephalopholis polleni (Bleeker, 1868)

ROCKCODS • SERRANIDAE

White-square rockcod

L: To 45 cm. Di: Indo-Pacific, throughout our area.
De: 10-150+ m. Ge: Usually only seen on deep clear water drop-offs, in some areas where common it also occurs on shallow reefs. It swims well out in the open and juveniles look like a basslet (below).

Gracila albomarginata (Fowler & Bean, 1930)

Coronation lyretail-cod
L: To 80 cm. Di: Indo-Pacific, throughout our area.
De: 3-200+ m. Ge: Juvenile with distinct colour pattern, changing gradually to all orange with small blue spots all over. Coastal shallow reef flats and outer reef crests and slopes. Juvenile (below) secretive in rock-boulder areas.

Variola louti (Forsskal, 1775)

White-edged lyretail-cod

L: To 65 cm. Di: Indo-Pacific, throughout our area, mainly equatorial waters. De: 3-100+ m. Ge: Coastal to outer reef crests, usually in rich coral growth areas. Adult has white posterior margin on the caudal fin and juveniles by the orange colour all over (below).

Variola albimarginata Baissac, 1952

ROCKCODS • SERRANIDAE

Barramundi cod

Length: To 65 cm.
Distribution: West Pacific, in all of our area, including Andaman Sea.
Depth: 1-40+ m.
General: Juveniles distinctly spotted, becoming grey to reddish brown as adults with proportionally smaller and more numerous spots. Coastal reefs and lagoons and deep, somewhat silty slopes to at least 40 m. Juveniles very small, about 30 mm long, from muddy bays on rocky outcrops.
A monotypic genus.

Chromileptes altivelis (Valenciennes, 1828)

Spotted coralcod
L: To 60 cm. Di: From Andaman sea throughout west Pacific. De: 3-40 m. Ge: A coastal species, preferring algae reef habitat in protected bays and lagoons. Often abundant on reefs just off shore in depths between 6-10 m. Spots in juveniles proportionally larger. The similar common *P. leopardus* below.

Plectropomus maculatus (Bloch, 1790)

Footballer coralcod
L: To 1 m. Di: Indo-Pacific, throughout our area.
De: 10-150+ m. Ge: A highly variable species, found in various habitats. Can change colour dramatically from shown distinct pattern to dark all over. Small juveniles mimic the poisonous pufferfish *Canthigaster valentini*.

Plectropomus laevis (Lacépède, 1801)

ROCKCODS • SERRANIDAE

Vermicular coralcod
L: To 70 cm. Di: Widespread Indonesia, ranging at least to Philippines. De: 3-50+ m.
Ge: The least observed species in the genus. Interestingly the juvenile (below) mimics *Cheilinus celebicus,* a common wrasse in our area. Adults are mainly found on deep outer reef drop-offs and seen solitair.

Plectropomus oligacanthus (Bleeker, 1854)

Purple grouper

Length: To 75 cm.
Distribution: West Pacific, throughout our area but replaced by next sp in Indian Ocean.
Depth: 2-150+ m.
General: Highly variable, small juvenile dark blue with orange tail, changing to light grey-blue with numerous tiny spots and somewhat larger ones scattered all over in large adults. Can have also yellow fins. Coastal reefs and large silty lagoons on outcrops with caves.

Epinephelus cyanopodus (Richardson, 1846)

Yellowfin grouper

Length: To 75 cm, reported to 90 cm.
Distribution: Andaman Sea to Bali, Indonesia.
Depth: 10-150 m.
General: Juveniles found in shallow waters, adults on deep reefs. The blue colouration of the body can be changed from a deep blue into a light blue. Older adults with white spots all over.

Epinephelus flavocaeruleus (Lacépède, 1802)

GROUPERS • SERRANIDAE

Long-finned cod
L: To 40 cm. Di: South-east Asia from Andaman Sea to southern Japan and Australia. De: 0-50+ m.
Ge: Pectoral fin large and ventral fins long, sometimes reaching anus. Shallow to deep coastal and estuarine reefs, juveniles sometimes in intertidal pools. Adults often seen resting. Similar and common **E. merra** below.

Epinephelus quoyanus (Valenciennes, 1830)

Coral rockcod

Length: To 50 cm.
Distribution: West Pacific, Indonesia and Malaysia to southern Japan, Fiji and eastern Australia.
Depth: 1-30 m.
General: Pale grey to brownish with large pale areas and small black spots all over. Juveniles with white- black-edged ocelli all over. Secretive coastal reefs, often silty or in rocky estuaries in crevices and large corals. A shy species, difficult to get close to.

Epinephelus corallicola (Valenciennes, 1828)

Marbled rockcod

Length: To 65 cm.
Distribution: Reported from Java, Malaysia, to southern Japan, Samoa and to tropical eastern Australia.
Depth: 2-100 m.
General: Juveniles jet-black with white blotches and spots, becoming small spotted with alternating dark and light saddles along back. Mostly in large lagoons with large coral outcrops. Juveniles in protected muddy bays or estuaries with rocky or coral outcrops.

Epinephelus maculatus (Bloch, 1790)

GROUPERS • SERRANIDAE

Epinephelus caeruleopunctatus (Bloch, 1790)

Small-spotted grouper

Length: To 60 cm.
Distribution: Indo-Pacific from broadly our area to east Africa and central Pacific.
Depth: 1-50 m.
General: Highly variable pale to dark-grey or brown with scattered white spot of various sizes and dark blotches. Secretive in shallow clear water reefs in protected coastal waters but also adults can be seen in caves along deep drop-offs, whilst juveniles can be found in rockpools.

Epinephelus areolatus (Forsskal, 1775)

Squaretail grouper

Length: To 40 cm.
Distribution: Indo-Pacific from broadly our area to east Africa and central Pacific.
Depth: 6-200 m.
General: A common species in muddy habitats, mostly with isolated rocky or land-debris outcrops on coastal slopes, often in small groups.
The truncate caudal fin with its pale posterior margin readily identifies this species from other similarly blotched species.

Epinephelus bontoides (Bleeker, 1855)

Dusky grouper

Length: To 35 cm.
Distribution: Indonesia to Solomon Islands, north to Taiwan.
Depth: 1-30 m.
General: A little known species but commonly observed in Flores, Maumere where it lives in coastal muddy bays from very shallow sandflats aroung objects on sand to debris washed down deep slopes. Also seen commonly on Bali's north coast, shallow amongst the rocky boulders.

GROUPERS • SERRANIDAE

Foursaddle grouper

Length: To 35 cm.
Distribution: Indo-Pacific, broadly our area to east Africa and central Pacific.
Depth: 1-25 m.
General: Shallow clear coastal reefs, mainly with rocky boulders and sparse coral growth. A shy species, usually hiding during the day. Comes out on dusk, the specimen in the photograph, taken in Bali, came out every day at the exact time just before dark, and sitting in the exact same spot on the reef.

Epinephelus spilotoceps Schultz, 1953

Snout-spots grouper

L: To 65 cm. Di: Indo-Pacific, commonly throughout our area. De: 3-50 m. Ge: Coastal bays and protected inner reefs from shallow reef flats to deep lagoons and along the base of drop-offs. Best recognised by the two spots on the snout as shown in the photograph below.

Epinephelus polyphekadion (Bleeker, 1849)

White-speckled grouper

Length: To 35 cm.
Distribution: Widespread tropical Indo-Pacific from broadly our area to east Africa and to Marshall islands and Fiji.
Depth: 3-30 m.
General: Clear coastal bays and on mixed coral and rock substrates. Secretive along drop-offs but often at entrance of ledge or cave. Sometimes sits on large sponges.
Best recognised by the dark head and densely spotted body and fins.

Epinephelus ongus (Bloch, 1790)

GROUPERS • SERRANIDAE

Brown-spotted grouper

L: To 1 m. Di: Indo-Pacific from broadly our area to east Africa and central Pacific. De: 1-100+ m. Ge: Mainly lives in estuaries and coastal bays. Adults along the base of small drop-offs with caves or with isolated large outcrops of reef, shipwrecks, etc in deeper water.

Epinephelus coioides (Hamilton, 1822)

Red-barred rockcod

Length: To 36 cm.
Distribution: Indo-Pacific throughout our area.
Depth: 1-150+ m.
General: Often a very common species, occurring in loose groups over reefs. Seems to prefer rocky boulders with sparse coral growth and sponges. Coastal silty reef to clear outer reef lagoons and reef crests.
Highly variable species, banding sometimes barely visible. At least six different populations.

Epinephelus fasciatus (Forsskal, 1775)

Bleeker's grouper

L: To 65 cm. Di: Indo-Pacific, including Malaysia, Singapore, Indonesia, Philippines, Taiwan, and as far as Oman. De: 1-50 m. Ge: Coastal, usually silty bays from shallow flats as juveniles to deep along mud slopes. Variable from dark to pale with medium sized orange spots. Lower half of caudal fin darker than rest which is more obvious in pale individuals.

Right:
Schooling basslets
Pseudanthias squamipinnis
in Andaman Sea.

Epinephelus bleekeri (Vaillant, 1877)

BASSLETS • SERRANIDAE

Longfin perchlet

Length: To 35 mm.
Distribution: Indo-Pacific from broadly our area to east Africa.
Depth: 3-75 m.
General: One of the tiniest species in the family Serranidae. Very secretive on rich coral growth. Rarely noticed and could easily be mistaken for a juvenile hawkfish. The photograph is the first published of this species *in situ*. Clear protected shallow coastal reefs to deep rich rubble slopes and several similar species.

Plectranthias longimanus (Weber, 1913)

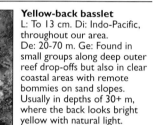

Yellow-back basslet
L: To 13 cm. Di: Indo-Pacific, throughout our area.
De: 20-70 m. Ge: Found in small groups along deep outer reef drop-offs but also in clear coastal areas with remote bommies on sand slopes. Usually in depths of 30+ m, where the back looks bright yellow with natural light.

Pseudanthias bicolor (Randall, 1979)

Randall's basslet

Length: To 7 cm.
Distribution: Indonesia, Philippines, Micronesia and southern Japan.
Depth: 15-70 m.
General: Mostly deep on outer reef drop-offs, but in some places relatively shallow. On Bali's north coast is occurs commonly along the top of some drop-offs in depths of 15 m.
Females plain compared to male with yellow on snout and tail, staying close to the substrate.

Pseudanthias randalli (Lubbock & Allen, 1978)

BASSLETS • SERRANIDAE

Two-spot basslet

Length: To 14 cm.
Distribution: Eastern Indian Ocean, ranging to Bali and Flores, Indonesia, however possibly more than one species.
Depth: 10-70 m.
General: Deep coastal reef slopes and along base of drop-offs. Usually in small aggregations with a few males dominating groups of females. Males highly variable from one area to the next and Indian and Pacific forms may represent different species (see photographs).

Pseudanthias bimaculatus (Smith, 1955)

Flores basslet

Length: To 14 cm.
Distribution: Only known from the Maumere area of Flores, Indonesia.
Depth: 1-50 m.
General: Coastal reef slopes and drop-offs in small groups dominated my a single male. Undetermined species, closely related to *P. bimaculatus* or a geographical variation. The females of the two species are virtually identical in colour.

Pseudanthias sp.

BASSLETS • SERRANIDAE

Pseudanthias fasciatus (Kamohara, 1954)

Red-stripe basslet

Length: To 15 cm.
Distribution: Indo-Pacific, throughout our area.
Depth: 20-150+ m.
General: Rare in depths less than 50 m in our area and only seen in the shallow part of its range in cooler water such as upwelling along southern Indonesia. Adults in large caves, often upside-down near the ceiling, along deep drop-offs. Juveniles in 20-30 m, usually in small aggregations. Was recently discovered off southern Africa by divers on mixed gas in large numbers

Pseudanthias luzonensis (Katayama & Masuda, 1983)

Luzon basslet

Length: To 12 cm.
Distribution: Philippines, Indonesia, New Guinea and northern Australia.
Depth: 20-60 m.
General: Rich coral and sponge reef slopes, usually protected sides of detached coastal reefs. Often in small loose groups, comprising mostly of females, swimming well above the substrate when feeding. Female similar to male but lacks the spot in the dorsal fin.

Pseudanthias hypselosoma (Bleeker, 1856)

Pink basslet

Length: To 10 cm.
Distribution: West Pacific, including all of our area, ranging to the Maldives.
Depth: 6-50 m.
General: Colour variable pale to darker pink, juveniles with red tips and narrow margin on caudal fin, and males changes during display in having a pale forehead and red caudal fin. Particularly common in Indonesia where they occur in large aggregations with coral heads.

BASSLETS • SERRANIDAE

Mirror basslet

Length: To 15 cm.
Distribution: Malaysia, Indonesia, Philippines to Samoa, Micronesia and Japan.
Depth: 10-150+ m.
General: Coastal to outer reef drop-offs. Usually very deep, over 30 m, but in some places in Indonesia at snorkel depth. Often in large groups of mixed sexes feeding well out in the open along drop-offs in currents on zooplankton. A highly variable species and blotch which typifies the male is sometime absent.

Pseudanthias pleurotaenia (Bleeker, 1857)

Red-band basslet
L: To 10 cm. Di: West Pacific, also in the Andaman Sea.
De: 10-60 m. Ge: Colour variable pale to darker pink, juveniles with red tips and narrow margin on caudal fin, and males develop a broad red band over the middle of the body. Congregates on isolated rocky outcrops along deep slopes.

Pseudanthias rubrizonatus (Randall, 1983)

BASSLETS • SERRANIDAE

Jewel basslet

Length: To 12 cm.
Distribution: Widespread throughout the Indo-Pacific, including all of our area but some geographical variations.
Depth: 3-20 m.
General: Reef crests, slopes and drop-offs, often in large aggregations, especially where the dominant species of the genus. Most reef habitats from silty coastal to clear outer reefs, along drop-offs and in lagoons. Colour highly variable from red, orange to purple and in some areas peppered with yellow spots. Photographes show variations.

Pseudanthias squamipinnis (Peters, 1855)

Red cheeked basslet

Length: To 9 cm.
Distribution: West Pacific, Indonesia, New Guinea to northern Australia and Philippines. Depth: 3-20 m.
General: A fairly common species, in small groups on reef crests and along upper region of drop-offs. Mainly rich coastal reefs with mixed stony and soft corals.
Males are easily recognised by the red stripe on their cheeks.

Right:
Schooling basslets with several mixed species in Manado, Indonesia.

Pseudanthias huchtii (Bleeker, 1856)

BASSLETS • SERRANIDAE

Pacific flame basslet
L: To 9 cm. Di: Malaysia, Singapore, Indonesia, Philippines. De: 1-15 m. Ge: Clear coastal to outer reefs in current prone areas. Usually occurs in large schools along the upper edge of drop-offs. Males often busy displaying to each other or to opposit sex by showing the red dorsal fin. Females below.

Pseudanthias dispar (Herre, 1955)

Indian flame basslet

Length: To 6 cm.
Distribution: Sri Lanka into the Andaman Sea.
Depth: 10-30 m.
General: Occurs in large aggregations on steep reef slopes in 20-30 m depth. Closely related and very similar to *P. dispar*, but has a red tail.

Pseudanthias ignitus (Randall & Lubbock, 1981)

Yellow tail basslet

L: To 10 cm Di: In our area only Indian Ocean: From Andaman Sea to Christmas Island. De: 5-40 m. Ge: More common in the western Indian Ocean. During daylight it moves over reef patches into the water column.

Pseudanthias evansi (Smith, 19554)

BASSLETS • SERRANIDAE

Lori's basslet

L: To 8 cm. Di: West Pacific, mainly Indonesia, ranging to Philippines and Australia. De: 15-70 m. Ge: Found mainly along clear water drop-offs in caves with black corals and large fans, but also in outer reef lagoons in rich coral areas.

Pseudanthias lori (Lubbock & Randall, 1976)

Purple queen

L: To 11 cm. Di: West Pacific, Indonesia to Philippines and Micronesia, and to Australia. De: 10-25 m. Ge: Clear coastal to outer reefs along drop-offs and slopes in current prone areas. May form large aggregations to feed well above the substrate. Females have a yellow stripe along base of dorsal fin.

Pseudanthias tuka (Herre & Montalban, 1927)

Sailfin queen

Length: To 15 cm.
Distribution: West Pacific, Indonesia, Philippines to Samoa, Micronesia, Japan and Australia.
Depth: 5-60 m.
General: Rare in Indonesia and only seen by the author north of Sulawesi, mainly from southern Japan ranging along eastern New Guinea to Australia and further east into Pacific.
Forms large schools on outer reef slopes, feeding high above the substrate.

Pseudanthias pascalus (Jordan & Tanaka, 1927)

SOAPFISHES • SERRANIDAE

Two banded soapfish

L: To 25 cm. Di: West Pacific, throughout our area. De: 1-50 m. Ge: Sometimes all black. Coastal reefs and quiet lagoons, enters estuaries and often with isolated outcrops. Feeds on a variety of invertebrates and fishes.

Diploprion bifasciatum Kuhl and van Hasselt, 1828

Lined soapfish

Length: To 25 cm.
Distribution: Indo-Pacific, throughout our area.
Depth: 1-150+ m.
General: Found in a variety of coastal habitats from shallow estuaries, lagoons, rockpools to deep muddy slopes with isolated outcrops of anything to provide shelted.
Usually young are observed in shallow depths. Large adults in which the lines brake up into dashes appear to be restricted to deep water.

Grammistes sexlineatus (Thunberg, 1792)

Snowflake soapfish
Length: To 34 cm.
Distribution: Tropical Indo-Pacific: All of our area, and to east Africa.
Depth: 15-150+ m.
General: A secretive species, not often seen but occurs commonly in selected areas, probably because of a specific habitat which is not clear to us. It seems to like small islands in current prone areas where it is found in loose groups in ledges or under corals with sand below. A distinct species with a curious beard-like skinflap on the chin and a distinct spot on the tip of the snout.

See photo on opposite page.

Pogonoperca punctata (Valenciennes, 1830)

SOAPFISHES • SERRANIDAE

Arrow-head soapfish

Length: To 15 cm.
Distribution: Indo-Pacific, all of our area to, east Africa, Micronesia, Japan and Australia.
Depth: 10-50 m.
General: Typically found in caves along drop-offs. A solitary species found commonly on coastal as well as outer reefs but with a preference for dark caves or ledges it is often missed by divers, except when using a torch. Easily identified by shape and colour.

Belonoperca chabanaudi Fowler & Bean, 1930

Yellow reef-basslet

Length: To 8 cm.
Distribution: Indonesia and Philippines to Coral Sea.
Depth: 20-50 m.
General: Found in the back of large and deep caves with rich invertebrate growth along steep outer reef drop-offs. Very secretive and difficult to photograph as they are often in areas too dark to focus and disturbed when using a light.

Liopropoma multilineatum Randall & Taylor, 1988

Swallow-tail seaperch
L: To 8 cm.
Di: West Pacific, Indonesia, Philippines to Micronesia, Japan and Australia.
De: 15-70 m. Ge: Found in caves and on ceilings of large overhangs, often swimming upside down, along outer reef drop-offs and small off shore islands. Ususually seen singly in Indonesia and Philippines but seen in small aggregations in Japan.

**Right: A pair of curious lyre-tail dottybacks *Pseudochromis steenei*, showing their dog-like teeth.
The sexes have different colours, the male has the orange head.**

Serranocirrhitus latus Watanabe, 1949

DOTTYBACKS • PSEUDOCHROMIDAE

A large family of small and often colourful fishes. Presently at least 6 genera and over 70 species are recognised, but this complicated family of small and secretive members is under revision with many new species recently discovered in the south-east Asian region. Most species are placed in Pseudochromis, but is comprises several distinct groups and is only provisionally used for some of the species included here, as they are waiting to be assigned to new genera in the family revision.

Dottybacks are commonly found on all kinds of reefs with numerous small hiding places such as cracks in rock faces, boulders, small caves or the old coralskeletons overgrown by new, and in mixed algae, sponge and coral habitats. Some species prefer silty coastal reefs and others exclusively live along prestine outer reef walls. Most species are seen solitair but their companion is probably out of sight. Some species live in small aggregations of mixed sizes and may have a large cave as a territory, occupying the narrow ledges and numerous narrow passages which are typical in coral reef drop-offs. They are small hunters, looking for almost anything which moves, including most crustaceans and small fishes. Some of the larger species can be very aggresive and are not suited for a community aquarium.

Some species are known to produce clusters of eggs which are guarded by the male. Hatchlings are about 2.5 mm long and postlarvae settle when about 15 mm long. Juveniles are less colourful than adults and in some species the sexes have different colours. The colour variations within one species can be unbelievable, changing with habitat and also geographically, to such extend that one would regard them as different species.

Identification of most species is easy from the photographs. The species included here are distinct and when sexes differ, both are shown. However, some dottybacks or rockbasslets are extremely variable, including our most common species P. fuscus which can be plain grey or brown, sometimes yellow over back, and sometimes has a totally bright yellow phase. Such variation can be expected in other, perhaps not as common, species and descriptions of new species from some areas are still suspect.

Slender dottyback

Length: To 12 cm.
Distribution: Indonesia, Philippines and northern Australia.
Depth: 3-30 m.
General: Coastal to outer reef slopes and drop-offs with dense coral and sponge growth, moving through passages and staying very close to substrate. Little variation in colour between various sizes or different sex, the white band easily identifies this species.

Pseudochromis bitaeniatus Fowler, 1931

DOTTYBACKS • PSEUDOCHROMIDAE

Magenta dottyback

Length: To 60 mm.
Distribution: Philippines to Samoa and southern Japan.
Depth: 6-65 m.
General: A uniformly but bright coloured species which wants to be noticed.
It swim out in the open but close to the substrate. Because the water filters out the red, it looks blue with the natural light. Occurs on outer reef walls and often on the sides of channels near ledges and small caves.

Pseudochromis porphyreus Lubbock & Goldman, 1974

Purpleback dottyback

Length: To 60 mm.
Distribution: Malaysia, Philippines
Depth: 5-25 m.
General: Lives in small groups among corals and rocks on reef slopes and bases of small drop-offs. A very shy species, but the most attractive dottyback in this region.

Pseudochromis diadema Lubbock & Randall, 1978

Two-tone dottyback

Length: To 50 mm.
Distribution: Indonesia, New Guinea, and northern Australia to Melanesia.
Depth: 10-50 m.
General: An easily identified species although Australian population is slightly different. In some areas commonly found along sponge and coral rich drop-offs along ledges and in small caves, ranging from coastal to outer reef habitat.

Pseudochromis paccagnellae Axelrod, 1973

DOTTYBACKS • PSEUDOCHROMIDAE

Dusky dottyback

Length: To 10 cm.
Distribution: Indo-Pacific, throughout our area, ranging to India, Taiwan and northern Australia.
Depth: 1-30 m.
General: A common and one of the most variable species, ranging from drab grey or brown to bright yellow. Sometimes yellow over the back. Almost every reef habitat from shallow coastal lagoons and seagrass beds with reef patches to outer reef lagoons.

Pseudochromis fuscus Müller & Troschel, 1849

Long-finned dottyback

Length: To 12 cm.
Distribution: Indonesia from Sulawesi north, Philippines Palau.
Depth: 10-30 m.
General: Along clear coastal and outer reefs with lots of sponge growth, especially long grey tubular species is which they often dissappear out of sight.
Swims in an unusualy way by strongly twisting its tail with short dashes through the passages which makes it almost impossible to get a good side-on picture.

Pseudochromis polynemus Fowler, 1931

Lyretail dottyback

Length: To 12 cm.
Distribution: Indonesia and northern Australia.
Depth: 8-50 m.
General: This species was recently described as distinct from *P. moorei* on the basis of colour, from a few specimens from Bali. However in Bali alone it is highly variable with males partly to completely orange, sometimes yellow tail, and females dark with yellow or dark tail similar to male. At Bali on deep sand slope with rubble ridges and small coral or rock outcrops with crinoids or with soft corals.

Pseudochromis steenei Gill & Randall, 1993

DOTTYBACKS • PSEUDOCHROMIDAE

Splendid dottyback

Length: To 13 cm.
Distribution: Philippines, eastern Indonesia and northern Australia.
Depth: 3-35 m.
General: A common species around the lesser Sunda Islands and in the Banda Sea, Indonesia, where they occur on rich coastal inner reef slopes with rich coral and sponge growth. Tiny juveniles mostly yellow are often in small groups in stringy-bushy looking sponges which provide good shelter. Adults often in large yellow tubular sponges, like *Thalysias vulpina* and *Clathria basilana*.

Pseudochromis splendens Fowler, 1931

Firetail devil

Length: To 20 cm.
Distribution: Malaysia, Indonesia, Philippines and southern Japan.
Depth: 1-20 m.
General: A variable species from greenish grey to bright red, more or less barred on the sides. Amongs channels on reef crests among the overgrown coral-rock which are covered with short algae and living coral colonies on top. Coastal as well as outer reefs, semi-exposed to surge, and tidal currents. Female shown, male has red tail.

Labracinus cyclophthalmus (Müller & Troschel, 1849)

LONGFINS • PLESIOPIDAE

A small family comprising six genera and about 20 species. Except for some large and spectacular temperate species, they are small and very secretive. Only two, but very interesting genera are included here: the Comet (genus *Calloplesiops*) is the best known and occasionally observed by divers in our area, and the devil fishes (genus *Assessor*). The first is also a popular aquarium fish which only recently has been bred in captivity. We show all stages of the Comet from juvenile to adult and even eggs and the larvae. This beautiful miracle basslet, as it is named in Europe, is a tropical species and also to be found in the Western Indian Ocean and the Red Sea.

There are three devilfishes in the genus *Assessor* which are small, living in caves, usually in small groups and typically swimming upside-down on the ceiling there. These little fishes, less than 10 cm long, are mouth breeders and further studies may well show that they belong in their own family. All other plesiopids or longfins are thought to nest amongst rocks or small crevices under the corals in which the male guards the eggs. The male *Assessor* incubates the eggs and they readily spawn in aquaria. Broods are produced on a monthly basis during the spawning season but hatchlings are tiny, only a few millimeter long and no attempts were made to raise the young.

Other species in this family in our area belong to the genus *Plesiops*, are small and mostly drab. Some are intertidal and not included here.

Yellow devil fish
L: To 6 cm. Di: Eastern Irian Jaya to northern Australia. De: 2-25 m. Ge: Protected reefs to high surge zones where they shelter in large open caves, usually upside down hoovering just below the ceilings. Easily recognised by the yellow colour and forked tail. Replaced by the grey-blue *Assessor randalli* in the Philippines and north to Japan, see photo below.

Assessor flavissimus Allen & Kuiter, 1976

LONGFINS • PLESIOPIDAE

Comet

Length: To 20 cm.
Distribution: Widespread tropical Indo-Pacific, including all of our area.
Depth: 3-50 m.
General: Number of spots increases with age and become proportionally smaller. In large adults the fins spots elongate, eventually join and form lines from the outer margin first. When scientists realized this, the similar species *C. argus* became a synonym. Although a common species it is not often seen unless one uses a torch when diving along drop-offs. They live among old coastal collapsed reef in muddy areas as well as outer reefs deep in the back of caves and ledges. Usually only one is visible but as many as three were known to live together in a small part of a reef less than one square metre at Flores.
In Japan several pairs were observed in a shallow reef with long overhangs.
All colourations of this beautiful species are shown right.
Eggs and postlarvae below.

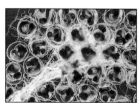

Calloplesiops altivelis (Steindachner, 1903)

BIGEYES • PRIACANTHIDAE

A small family of distinctly looking fishes, comprising 4 genera and about 17 species, distributed world wide in tropical to subtropical seas. They are known as red bullseyes, goggle-eyes and glass-eyes, because of their most obvious feature: the very large eye. The head is large and a prominent spine at angle of preopercle. The mouth is very oblique, near vertical and ventral fins connected by a membrane to belly. Nocturnal fishes, usually reddish brown to deep red depending on depth. Feeds on crustaceans, cephalopods and fishes. Generally solitair, but known to school in some areas. They can be found on silty coastal reefs or with outcrops on sand and protected areas along outer reefs or venture to very deep waters.

Despite their very large mouth they prefer tiny prey, mostly taken at night when they move into the open water colomn to feed on zooplankton, especially small shrimps or mysids. Juveniles make excellent aquarium fish, though when fed regularily they grow fast.

Glasseye

Length: To 30 cm.
Distribution: Tropical Indo-Pacific, ranging throughout our area.
Depth: 3-30 m.
General: Anal fin from rounded in young to truncate in adults. Colour usually bright red, when hiding or at night, silvery when in the open during the day, and usually a series of dark spots along the lateral line. Lagoons and protected outer reefs. The Crescent-tail bigeye, *P. hamrur,* is similar except for the shape of the tail and also common in our area.

Priacanthus blochii Bleeker, 1853

Blotched bigeye

Length: To 35 cm.
Distribution: All tropical seas, throughout our area.
Depth: 3-30 m.
General: One of the most widespread species. At night it is seen floating just above the reef, virtually motionless like keeping still to hear the softest noise, its way to sense prey in the dark.

Right: Estuaries and mangroves are the home of many cardinalfishes like *S. orbicularis.*

Heteropriacanthus cruentatus (Lacépède, 1801)

CARDINALFISHES • APOGONIDAE

The cardinalfishes are a very large family of small reef fishes, most numerous in tropical waters. They have a global distribution with 3 subfamilies and 26 genera containing an estimated 250 species. Many species in the range of this book were recently discovered and are new to science. The majority of species live in shallow coastal waters, occupying caves and crevices during the day and drifting out in the open during the night to feed. A few species live deep on continental slopes and some are restricted to freshwater. Their diets consist primarily of small planktonic invertebrates but a few feed on small fishes as well. Apogons typically have 2 spines in the anal fin and two usually separate dorsal fins. The reef dwellers are distinctly marked with stripes and spots.

Most species are placed in the genus *Apogon* which includes most of the striped species, however, another genus with several species in our area, *Cheilodipterus,* has stripes, too, but these have large canine teeth. A large number of striped *Apogon* are closely related and comprises several sibling species and geographical variations. The status od many is not clear. E. g. *Apogon cyanosoma,* a pale species with thin orange lines. Has several closely related species and all are usually referred to the same ‚catch-all‘ name. Some geographical variations are distinct and may represent good species. The cardinalfishes from the area of this book are presently under investigation by ichthyologists from Australia and Japan.

Apogonids are one of the few marine fish families which use oral brooding. After fertilisation the male takes the eggs in the mouth for incubation which lasts just over one week in tropical waters and slightly longer in cooler zones. Reports of females incubating eggs appears to be based on males actually eating a brood in part, or whole before taking on a second brood, thus appeared to be gravid during incubation. The hatching larvae have a long planktonic stage, lasting about 60 days, which can disperse them over a great distance. Apart from brooding males the sexes are very similar. In some species the females are larger and more colourfull, particularily during courtship in which she dominates. Most species pair when adult, including many of those in schools which is not always obvious without close examination.

Nose-spot cardinalfish
L: To 60 mm. Di: Widespread tropical Indo-Pacific, including all of our area. De: 10-50 m. Ge: Diagnostic markings are the black elongated spot on the nose and dusky outer caudal fin rays. During the day usually found along drop-offs with large fans, and sometimes forming massive dense schools.

Rhabdamia cypselura Weber, 1909

CARDINALFISHES • APOGONIDAE

Tiger cardinalfish

L: To 22 cm. Di: Indo-Pacific, including all of our area.
De: 3-30 m. Ge: Coastal reefs and rocky estuaries. Best separated from similar species by yellow on head of which often traces remain in adults, particularly in the iris. Juvenile below.

Cheilodipterus macrodon Lacépède, 1802

Eight-line cardinalfish

Length: To 24 cm.
Distribution: Malaysia, Indonesia and Papua New Guinea.
Depth: 3-20 m.
General: Distinctly black-tipped on 1st dorsal fin and black margins along 1st soft rays and outer caudal fin. Solitair along shallow coastal drop-offs, coming out on dusk. Common in equatorial waters but probably widespread. The largest species in the genus.

Cheilodipterus alleni Gon, 1994

Five-line cardinalfish

Length: To 10 cm.
Distribution: Widespread tropical Indo-Pacific, including all of our area, ranging into warm-temperate zones.
Depth: 3-40 m.
General: Distinctly marked, juveniles more yellow on caudal peduncle, but almost identical to *C. isostigma* from which it differs in lacking anterior canines in lower jaw and position of caudal peduncle spot lining up with preceding black stripe. Coastal reef crests and slopes to outer reefs.

Cheilodipterus quinquelineatus Cuvier, 1828

CARDINALFISHES • APOGONIDAE

Apogon aureus (Lacépède, 1802)

Ring-tail cardinalfish

Length: To 14 cm.
Distribution: Widespread tropical Indo-Pacific, including all of our area.
Depth: 1-50 m.
General: Adults distincly marked with black band on caudal peduncle, which in small juveniles develops as a large spot. Coastal reef crests and slopes, usually in small to large aggregations near large overhangs. Adults pair within aggregations.

Apogon bandanensis Bleeker, 1854

Banda cardinalfish

Length: To 10 cm.
Distribution: Widespread throughout our area, ranging to Samoa.
Depth: 3-30 m.
General: One of three similar species. This species has yellow margins on caudal fin and on the soft dorsal and anal fin tips, and banding on body more defined in adults.
Coastal lagoons and reefs, usually silty areas.

Apogon dispar Fraser and Randall, 1976

White-spot cardinalfish

Length: To 65 mm.
Distribution: Widespread tropical west Pacific, including all of our area.
Depth: 10-60 m.
General: Body semi-transparent, a distinct white spot immediately above a black spot on caudal fin base. Usually in small aggregations with black corals along drop-offs in large overhangs and caves.

CARDINALFISHES • APOGONIDAE

False three-spot cardinalfish

Length: To 14 cm.
Distribution: Indonesia and New Guinea, probably wide-ranging.
Depth: 3-30 m.
General: Very similar in behaviour as well as looks to *A. trimaculatus* but paler with scales thinly dark edged and no spot on opercle. Secretive on clean coastal reefs, coming out on dusk.

Apogon rhodopterus Bleeker, 1854

Cheek-bar cardinalfish

Length: To 10 cm.
Distribution: Indonesia and Philippines.
Depth: 3-25 m.
General: Similar to *A. chrysopomus* but spots replaced by bars on opercle and spots absent on cheeks. Found in same area but schools among branching corals and seems to prefer clearer waters though still on coastal reefs and in lagoons.

Apogon sealei Fowler, 1918

Cheek-spots cardinalfish

Length: To 11 cm.
Distribution: Indonesia and Philippines.
Depth: 3-10 m.
General: Two dark body lines and orange spots on opercle and cheek. Some geographical variations in which body lines are replaced by spots and general colour differences. Indonesia and Philippines. Quiet and protected coastal reefs. Usually very shallow, in pairs over corals.

Apogon chrysopomus Bleeker, 1854

CARDINALFISHES • APOGONIDAE

Apogon kiensis Jordan & Snyder, 1901

Rifle cardinalfish

Length: To 8 cm.
Distribution: Widespread tropical Indo-Pacific, including all of our area, ranging into warm-temperate zones.
Depth: 6-50 m.
General: Dusky area posteriorly on anal fin base. Distincly striped; broad black middle stripe edged in white from snout to caudal fin margin readily identifies this species. In small aggregations in lagoons and estuaries, on deep sand and mud slopes.

Apogon nigrofasciatus Lachner, 1953

Black-striped cardinalfish

L: To 10 cm. Di: Widespread tropical Indo-Pacific, including all of our area. De: 3-50 m. Ge: Coastal to outer reef drop-offs and rocky estuaries, usually in small aggregations in crevices. The similar and also common **A. angustatus** below.

Apogon cyanosoma Bleeker, 1853

Orange-lined cardinalfish

L: To 9 cm. Di: Indo-Pacific, including all of our area. De: 1-35 m. Ge: Several similar species confused under this name. Below: **A. properuptus** from Southern Indonesia (Java), New Guinea and Australia.

CARDINALFISHES • APOGONIDAE

Multi-striped cardinalfish

Length: To 10 cm.
Distribution: Widespread tropical west Pacific, ranging to east Indian Ocean, including all of our area.
Depth: 3-25 m.
General: Numerous thin black lines and black face with thin pale lines identifies this species. Alternating thin and thicker lines in adults which is more pronounced in juveniles. Coastal reef flats and slopes with bommies, usually in small numbers.

Apogon multilineatus Bleeker, 1853

Moluccen cardinalfish

Length: To 85 cm.
Distribution: Widespread west Pacific, from Australia, through eastern Indonesia and New Guinea to Philippines.
Depth: 3-25 m.
General: Juveniles variable, usually pale with a wide orange-brown stripe, and adults plain and brown to dark brown. White spot following base of 2nd dorsal fin usually very distinct.
Coastal reefs and lagoons.

Apogon moluccensis Valenciennes, 1828

Ruby cardinalfish

Length: To 50 mm.
Distribution: Widespread tropical Indo-Pacific, including all of our area.
Depth: 3-55 m.
General: Reddish with dusky centred scales. Similar species with longer caudal peduncle or taller first dorsal fin may occur in Indonesian waters. The most common of several secretive species, only observed at night. Coastal reefs along drop-offs.

Apogon erythrinus Snyder, 1904

CARDINALFISHES • APOGONIDAE

Pyjama cardinalfish
L: To 85 mm. Di: Indo-Pacific, including all of our area. De: 3-25 m. Ge: Inshore and lagoons in aggregations amongst dense branching or large staghorn corals, in small to very large aggregations. Juvenile below.

Sphaeramia nematoptera (Bleeker, 1856)

Polkadot cardinalfish
Length: To 10 cm.
Distribution: Widespread tropical Indo-Pacific, including all of our area.
Depth: Surface zone.
General: Plain greyish with black bar and spots. Inshore, mangroves and lagoons nearby. Schooling and mainly in surface waters, commonly found under jetties and in harbours in the shade of pilons or decking.

Sphaeramia orbicularis (Cuvier, 1828)

Urchin cardinalfish
Length: To 20 mm.
Distribution: Indo-Pacific, throughout our area.
Depth: 1-35 m.
General: Variable, almost black with small white spots to striped pattern. Mainly with *Diadema* urchins, sheltering at the bases of the long spines. Once observed with a ball of juvenile *Plotosus* catfish.

Siphamia versicolor (Smith & Radcliffe, 1911)

TILEFISHES • MALACANTHIDAE

A small tropical family with representatives globally, comprising 2 genera and 11 species. The genus *Hoplolatilus* with the majority of species is confined to tropical coral reefs. The 2 Indo-Pacific species of *Malacanthus* are widespread and ranging into warm-temperate zones, and are commonly known as blanquillos. These slender fishes are found on sandy reef flats, either solitair or in pairs, swimming typically just above the bottom and dashing over short distances for a quick stop to study surroundings. Burrows are made under rocks on sand and often several are in the vicinity of their feeding range, to quickly go for cover on a potential threat. Adult may build large nesting sites, shifting large quantities of sand. They are distinctly marked with stripes on the body or tail, juveniles differ in some species greatly from adults. Eggs are probably laid in their large mounts, built of rubble, sometimes measuring over one metre in diameter, guarded by a pair. Hatchlings are pelagic, and postlarval stages are large when settling, 3-7 cm, suggesting a prolonged pelagic life.

Flagtail blanquillo

Length: To 30 cm, usually to 20 cm.
Distribution; Widespread tropical Indo-Pacific, ranging to east Pacific.
Depth: 6-50+ m.
General: Caudal fin rounded in juveniles becoming more truncate in adults. Little variation in colour from juvenile to adult. Settling young very pale, near translucent with dark oblique marks in fins. Quiet coastal reefs along its margins on sand or in sand flats within large reefs. Settling juveniles are large, 7-9 cm, secretive under rocks. Feeds primarily on sand dwelling invertebrates.

Malacanthus brevirostris Guichenot, 1848

Blue blanquillo

Length: To 40 cm.
Distribution: Indo-Pacific, throughout our area.
Depth: 6-40 m.
General: Juvenile (below) very different from adults. Adults usually in pairs on outer reef crests, juveniles on coastal reef slopes, often quite deep.

Malacanthus latovittatus (Lacépède, 1801)

TILEFISHES • MALACANTHIDAE

Hoplolatilus luteus Allen & Kuiter, 1989

Golden tilefish
Length: To 10 cm.
Distribution: Flores, Indonesia.
Depth: 15-35 m.
General: Only known from an unique coastal area at Flores where several other new fishes were discovered, including a wrasse not known elsewhere, and some fishes considered extremely rare in general live there commonly. The golden filefish was found on the muddy bottom of the channel, about 50 m wide, between the reef and the shore where they paired, quickly retreating to borrows when approached. The channel's depth is 35 m, and one pair was seen on a mud bank nearby at 15 m depth.

Hoplolatilus fourmanoiri Smith, 1963

Yellow-spotted tilefish

Length: 14 cm.
Distribution: Philippines, Vietnam.
Depth: 5-70 m.
General: Described from specimens found off Vietnam, and until now it is only recorded elsewhere from the Philippines. Prefers silty sand bottoms with coral rocks.

Hoplolatilus marcosi Burgess, 1978

Red-stripe tilefish

Length: 12 cm.
Distribution: Philippines, Northern Indonesia to Solomon Islands.
Depth: 30-80 m.
General: A pairing deep dwelling species on sandy and coral rubble areas. Shy and difficult to take photos of. Built great mounts for nesting sites.

TILEFISHES • MALACANTHIDAE

Purple tilefish

Length: 15 cm.
Distribution: Philippines.
Depth: 20-70 m.
General: A secretive species living on sandy and coral rubble. Swims usually in pairs: one partner of these plankton feeders swims some meters above their burrow while the other moves only around the entrance.

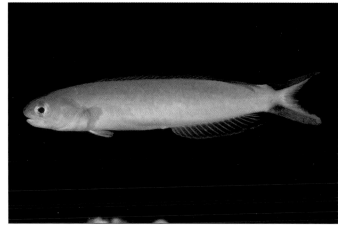

Hoplolatilus purpureus Burgess, 1978

Chameleon tilefish

Length: 11 cm.
Distribution: Philippines.
Depth: 20-40 m
General: Pairwise swimming species. Known for its ability to change its colour quickly from light green to dark blue.

Hoplolatilus chlupatyi Klausewitz et al., 1978

Blue-head tilefish
L: To 15 cm.
Di; Malaysia, Indonesia to east of our area.
De: 20-50+ m. Ge: Along the slopes on the bases of drop-offs. In most areas rarely seen in depths less than 30 m. Coastal as well as outer reefs, in pairs and stays close to burrows. Juvenile below.

Hoplolatilus starcki Randall & Dooley, 1974

129

COBIA • RACHYCENTRIDAE

Cobia

Length: To 2 m.
Distribution: All tropical seas, except east Pacific.
Depth: 2-50 m.
General: Juveniles distinctly striped, becoming plain brownish in large adults. Juveniles usually singly, adults usually sighted in small groups of 3-6 individuals, entering shallow water. This family contains a single wide-ranging species. A large pelagic species occasionally visiting reefs and under water have the appearance of a shark, which can be scary for a diver(!) as these fishes are curious and often come quickly and in close range.

Rachycentron canadum (Linnaeus, 1766)

REMORAS • ECHENEIDIDAE

A small distinct family, easily recognised by the large sucking disc dorsally on the head and nape, comprising 4 genera and 8 species. The sucker disc in fact is a modified first dorsal fin, comprising of transverse movable laminae, and is used to attach by suction to other fish, usually sharks or rays. Some are host-specific, however others may attach to about anything which moves, including ships and divers. Food is obtained when the host is feeding, but parasitic copepods which may attach to the host are taken as well. They feature elongate bodys with tiny embedded scales, a depressed head and the lower jaw greatly protruding, and dorsal and anal fin similar and mirror-like opposite.

Included here are the most commonly observed species. Other remoras are similar but usually lack stripes and are more stocky in body shape, except for a small but very rare species which is even more slender. This species only grows to 45 cm and has a very small head, usually found on large barracuda: Please keep an eye out for it.

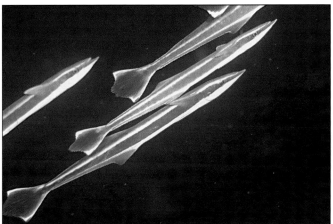

Slender suckerfish
L: To 1 m. Di: All tropical seas, except eastern Pacific. De: Surface to very deep, depending on host. Ge: Mostly pelagic off-shore, but enters coastal and estuarine waters with host. Usually attached to sharks and rays, often swimming free. Also known as Shark remora.

Echeneis naucrates Linnaeus, 1758

TREVALLIES • CARANGIDAE

A large tropical to temperate family, comprising approximately 25 genera and 140 species globally. Many are pelagic and wide-ranging, some enter brackish water. Many of the post-larval stages swim with floating objects or weed rafts, taken well beyond their normal range. The trevallies are streamlined fast swimming, usually schooling, fishes, with some diversity in shape and sizes. Carnivorous, smaller species feeding on zoo-plankton, and larger species feeding on other fishes such as pilchards, mullet or herring type species.

The largest species are tuna-like, growing to 2 m long and a massive weight of 80 kg, and swim mostly in schools off-shore. A few species are coastal and estuarine, sometimes only as juveniles, but many species hunt closely to reefs either singly or in small groups. A few small species are only found in protected coastal bays where they often form massive schools.

Golden trevally
Length: To 1 m.
Distribution: Widespread tropical Indo-Pacific, including all of our area.
Depth: Surface to 50+ m.
General: Usually in small groups. Juveniles (below) pelagic, under floating objects or swim close to large pelagics, as well as on coastal reefs.

Gnathanodon speciosus (Forsskal, 1775)

Highfin amberjack
L: To 1 m.
Di: All tropical to sub-tropical regions. De: Surface to 140 m.
Ge: Distinct oblique dark stripe over head through eye, becoming indistinct in some large adults. Young (below) under floating weed-rafts.

Seriola rivoliana Valenciennes, 1833

TREVALLIES • CARANGIDAE

Big-eye trevally

Length: To 85 cm.
Distribution: Indo-Pacific, containing all of our area.
Depth: 3-100 m.
General: A very common species in our area, often in great schools **(see page 131)**. Small juveniles with dark cross bars. Large adults singly or schools along deep drop-off, usually deeper than 30 m, like Sipadan, Malaysia and Banda Sea, Indonesia. Juveniles in estuaries and forming schools in coastal bays as they grow and move out to deeper water.

Caranx sexfasciatus Quoy and Gaimard, 1825

Giant trevally

Length: To 1.7 m.
Distribution: Widespread tropical Indo-Pacific.
Depth: 1-50+ m.
General: Dusky grey above and silvery below, juveniles with yellowish anal and lower caudal fin. Juveniles shallow in coastal bays, adults on deep slopes and inner reefs, usually deep in 20+ m, on sand flats and in reef channels.
The largest trevally attaining a weight of 62 kg (catch record).

Caranx ignobilis (Forsskal, 1775)

Blue-fin trevally

Length: To 70 cm.
Distribution: Widespread tropical Indo-Pacific, containing our area.
Depth: 6-50 m.
General: Greyish above, sometimes faintly banded, median fins blue, and small dark spots over body. Common on coastal reef slopes, often solitair following snapper which are benthic feeders to snap-up disturbed prey. Sometimes in small groups and occasionally seen in large schools during the spawning season.

Caranx melampygus Cuvier, 1833

TREVALLIES • CARANGIDAE

Banded trevally

Length: To 70 cm.
Distribution: Widespread tropical Indo-Pacific, containing all of our area.
Depth: 10-60 m.
General: Juveniles (below) often seen swimming with jellies. Coastal to outer reefs solitair or schools.

Carangoides ferdau (Forsskal, 1775)

Thicklip trevally

Length: To 60 cm.
Distribution: Indo-Pacific, throughout our area.
Depth: 3-50+ m.
General: Juveniles yellowish, similar to *C. ferdau* but longer fins. Adults bluish-grey with blue fins, some scattered dark spots and a few yellow spots on body. Juveniles in quiet coastal bays, along the bases of reef, singly or small groups. Adults deep in reef channels.

Carangoides orthogrammus (Jordan and Gilbert, 1882)

Yellow-spotted trevally

Length: To 45 cm.
Distribution: Indo-Pacific, including all of our area.
Depth: 3-50 m.
General: All golden or dusky banded with numerous small yellow spots all over body. Juveniles solitair on coastal, often silty reefs. Adults usually in small groups along deep drop-offs, swimming close to walls.

Carangoides bajad (Forsskal, 1775)

TREVALLIES • CARANGIDAE

Coachwhip trevally

Length: To 45 cm.
Distribution: Indo-Pacific, including all of our area.
Depth: 6-35 m.
General: Mainly on coastal mud flats in protected bays, swimming near the bottom in small groups, hunting small fishes and gobies in particular. Easily identified by the long filamentous dorsal fin with a black spot at the base of the filament part.

Carangoides oblongus (Cuvier, 1833)

Bludger trevally

Length: To 90 cm.
Distribution: Indo-Pacific, including all of our area.
Depth: 6-50 m.
General: Young mainly in coastal, often muddy, bays, feeding along upper parts of slopes or flat parts adjacent to it. Adults move to deeper water near off-shore reefs. In small groups with largest leading and smallest last.

Carangoides gymnostethus (Cuvier, 1833)

Bar-cheek trevally

Length: To 45 cm.
Distribution: Indo-Pacific, including all of our area.
Depth: 10-100+ m.
General: Fairly common in our area, swims usually solitair along steep drop-offs from coastal to outer reefs. Often around shipwrecks.
Best recognised by small dark bar on cheek which none of our other species has.

Carangoides plagiotaenia (Bleeker, 1857)

TREVALLIES • CARANGIDAE

Snub-nose dart

Length: To 65 cm.
Distribution: Indo-Pacific, including all of our area.
Depth: 0.5-50 m.
General: Silvery, bluish-green above, and lower anterior parts yellowish. Coastal reefs, bays and lagoons, usually in small to large aggregations. Juveniles in surface waters along beaches.
Adults school in deeper parts of protected bays and lagoons.

Trachinotus blochii (Lacépède, 1801)

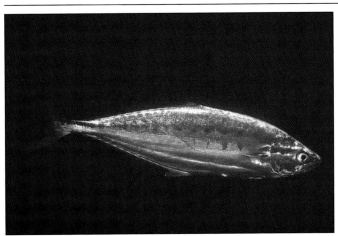

Queenfish

Length: To 70 cm.
Distribution: Indo-Pacific, including all of our area.
Depth: 3-100 m.
General: Silvery, bluish-green above, juveniles with a single row, adults with a double row of large dark blotches on sides. Coastal reefs, young enter estuaries, singly or small groups. Adults in various depths, occasionally seen very deep. Possesses venomous spines, particularly on the anal fin, which can cause a painfull wound.

Scomberoides lysan (Forsskal, 1775)

Yellowtail scad

Length: To 30 cm.
Distribution: Indo-Pacific, including all of our area.
Depth: 3-20 m.
General: A common species in coastal bays, mainly estuarine, schooling near mangroves or in small groups swimming fast along coastal sand or mud flats whilst feeding on zooplankton. Best recognised by the yellow tail and dusky bars on side which show strongly underwater.

Atule mate (Cuvier, 1833)

SNAPPERS • LUTJANIDAE

A large family comprising four subfamilies, 17 generas and 103 species world wide, most of which found in the west Pacific. The largest subfamily is Lutjaninae with 6 genera and 72 species, 43 of which in the west Pacific. Some doubt exists about the status of the family Caesionidae being only a subfamily of Lutjanidae. A total of 68 species occur in our area, many of which restricted to deep water and some occur mainly in estuaries and even freshwater. Included here are the reef dwellers which are commonly observed by scubadivers.

Snappers live in most benthic habitats and many species change their life style with growth, and often chance as juvenile from estuarine or freshwater to to adult in a full marine environment. They are pelagic spawners, producing a great number of tiny eggs less than 1 mm diameter. Postlarvae settle when about 20-40 mm long, depending on the species. All are carnivores, feeding on small fishes and invertebrates, either benthic or planktonic. Behaviour varies greatly in the family in the way they occur solitair or in schools between size and species.

Green jobfish

Length: To 100 cm.
Distribution: Indo-Pacific, widespread in our area.
Depth: 10-100 m.
General: It is a voracious hunter swimming high above substrate along slopes. Darts into the reefs hunting octopuses, cuttlefishes and also large fishes. Large individuals may be ciguatoxic. Very shy to divers.

Aprion virescens Valenciennes, 1830

Pinjalo snapper

Length: To 80 cm.
Distribution: West Pacific and Indian Ocean
Depth: 3-100 m.
General: In spite of a wide distribution this species is rarely obeserved by divers. Occurs in groups, generally very deep, but in some parts of Indian Ocean in shallow bays. Photo was taken in the Andaman Sea.

Pinjalo pinjalo (Bleeker, 1845)

SNAPPERS • LUTJANIDAE

Midnight snapper
L: To 60 cm. Di: Widespread in our area. De: 5-50 m.
Ge: Juveniles solitair, post-larvae with crinoids, adults may form loose aggregations. Coastal and inner reefs as well as on outer reef drop-offs. Juveniles have extended ventral fins which gradually reduce with growth.

Macolor macularis Fowler, 1931

Black snapper

Length: To 60 cm.
Distribution: Indo-Pacific, throughout our area.
Depth: 5-90 m.
General: Juveniles solitair on coastal reefs and in lagoons. Adults form schools along drop-offs on coastal and inner reefs.

Macolor niger (Forsskal, 1775)

Red emperor
Length: To 1 m, usually to 60 cm.
Distribution: Indo-Pacific, widespread in our area.
Depth: 2-100 m.
General: Small juveniles often seek protection in between spines of sea-urchins living on mud and sand slopes. Gathering urchins may bring juveniles together which form groups before leaving their host into deep water. Shown a large juvenile, its black bands turning brown and getting ready migrate to deeper water where as adult change to a uniform deep red, a stage which is mostly known from fishing.

Lutjanus sebae (Cuvier, 1828)

SNAPPERS • LUTJANIDAE

Humpback snapper

Length: To 50 cm.
Distribution: Indo-Pacific, throughout our area.
Depth: 1-30 m.
General: Juveniles usually in mangrove or seagrass areas in the intertidal zone. Adults form schools, often in great numbers, along reef slopes with little activity during the day. Very large individuals are sometimes solitair in shipwrecks. The colour varies greatly and adults are often red. Juveniles have similar colours to the adult shown.

Lutjanus gibbus (Forsskal, 1775)

Red bass

Length: To 75 cm.
Distribution: Indo-Pacific, throughout our area.
Depth: 3-70 m.
General: Juveniles on reef crests mimicing damselfish when small, hunting prey during the day, called two-spot snapper. Adults in moderate depths, usually solitair but sometimes in big schools. Taking this photo, some grey reef sharks were observed how they darted into the snappers to separate single fishes from the school. These were eaten at once.

Lutjanus bohar (Forsskal, 1775)

SNAPPERS • LUTJANIDAE

Timor snapper

Length: To 50 cm.
Distribution: Widespread tropical west Pacific.
Depth: 10-130 m.
General: Inhabits coastal and often muddy environments. Juveniles solitair or in small numbers with isolated outcrops of reefs or debris on sand and mud slopes in relatively shallow depths. Adults are red with pinkish bellies, and prefer deep water and rarely seen by scubadivers. Similar to next species, differing in the black and white spots on the caudal peduncle.

Lutjanus timorensis (Quoy & Gaimard, 1824)

Malabar snapper

Length: Can reach 1 m, but usually to 50 cm.
Distribution: Widespread tropical west Pacific.
Depth: 10-100 m.
General: Lives in muddy environments from coastal to off shore. Juveniles in the shallow part of the range, but adults only deep and normally not seen by divers. Adults mainly red and caught by lines and trawl.

Lutjanus malabaricus Schneider, 1801

Mangrove jack

L: To 1.8 m. Di: Indo-Pacific, throughout our area.
De: Intertidal to 100 m.
Ge: Juvenile (below) in mangroves and in lower reaches of freshwater systems. Larger juveniles school on coastal reefs and eventually move out to deeper water.

Lutjanus argentimaculatus (Forsskal, 1775)

SNAPPERS • LUTJANIDAE

Blacktail snapper

Length: To 40 cm.
Distribution: Widespread tropical Indo-Pacific, including remote islands.
Depth: Intertidal to 40 m.
General: Adults in 6-40 m, but usually deep in lagoons and coastal to outer reef slopes. Juveniles usually intertidal and in freshwater run-offs where they may form loose groups, they differ from the adult by lacking the black tail. Adults usually solitair.

Lutjanus fulvus (Schneider, 1801)

Lunartail snapper

Length: To 35 cm.
Distribution: From northern Indian Ocean to Philippines and to Vanuatu.
Depth: 10-50 m.
General: Mainly seen as adults on reefs, solitair or in small loose groups, near ledges and caves. Common in the Indian Ocean but in the Pacific part of our area rare.
A distinct species with the black moon shape pattern in the tail and the body bright yellow below.

Lutjanus lunulatus (Park, 1797)

One-spot snapper

Length: To 60 cm.
Distribution: Widespread tropical Indo-Pacific.
Depth: 5-30 m.
General: Adults occur in sheltered reefs with caves and are often with shipwrecks. Mostly solitair or in small loose groups which appears to be because sharing a suitable habitat, rather than schooling desire.
Juveniles are not known and perhaps overlooked because of similarities to other species or in contrast to most species occur in deep water.

Lutjanus monostigma (Cuvier, 1828)

SNAPPERS • LUTJANIDAE

Brown-stripe snapper

Length: To 40 cm.
Distribution: Indo-Pacific, throughout our area.
Depth: 10-40 m.
General: Adults solitair or in small groups, usually in deep lagoons or flat areas between reefs with coral bommies. Juveniles on coastal mud or sand slopes in small groups seeking protection from predation by sheltering near stinging anemones such as the tube anemones, *Cerianthus* spp., often sharing with other juvenile fish such as apogonids or *Amblypomacentrus*-damsels.

Lutjanus vitta (Quoy & Gaimard, 1824)

Ehrenberg's snapper

Length: To 35 cm, but rarely over 25 cm.
Distribution: Indo-Pacific, throughout our area.
Depth: 5-20 m.
General: A coastal species, usually schooling over shallow reefs near mangroves, in lagoons and sometimes congregate in large number near river outlets. Juveniles mainly in brackish mangove and estuarine zones. The photograph of the 15 cm individual was taken in brackish waters. Small juveniles have thicker and fewer orange lines.

Lutjanus ehrenbergii (Peters, 1869)

Russell's snapper

L: To 45 cm.
Di: Widespread tropical Indo-Pacific.
De: Intertidal to 80 m.
Ge: Juveniles in shallow protected bays and mangrove areas, entering lower reaches of freshwater ways. Adults in deep and usually off shore reefs. Two forms shown, sometimes faded stripes Juvenile similar but stripes are proportionally much thicker.

Lutjanus russelli (Bleeker, 1849)

SNAPPERS • LUTJANIDAE

Five-line snapper

Length: To 38 cm.
Distribution: Indo-Pacific, throughout our area.
Depth: 2-40 m.
General: Coastal reefs, lagoons and outer reef slopes. In small agregations or forms large schools. Juveniles in estuaries feeding in sand areas. They are similar to the adult but are less yellow and the dark spot shown in the photograph always distinct. Adults can turn-off the spot and are easily confused with the *L. kasmira*, but have five instead of four stripes.

Lutjanus quinquelineatus (Bloch, 1790)

Blue-striped snapper

L: To 35 cm. Di: Indo-Pacific. De: 5-265 m. Ge: A versatile species, occurring in shallow and very deep water in large schools. The very similar *L. bengalensis* is mainly an Indian species, but ranges east to Bali, swimming with *L. kasmira*. It is best recognised by the whitish lower third of the body.

Lutjanus kasmira (Forsskal, 1775)

Spanish flag snapper

Length: To 40 cm.
Distribution: Tropical west Pacific.
Depth: 2-80 m.
General: Coastal reefs and within sheltered areas of outer reefs. Usually in loose agregations on the top of reefs or along low drops. A rather shy species. Adults variable with lines from pale yellow to dark brown. Juveniles similar to adults with stripes more distinct and they occur in similar habitat as the adults but are more secretive in corals.

Lutjanus carponotatus (Richardson, 1842)

SNAPPERS • LUTJANIDAE

Lutjanus decussatus (Cuvier, 1828)

Checkered snapper

Length: To 30 cm.
Distribution: Tropical west Pacific and eastern Indian Ocean.
Depth: Intertidal to 30 m.
General: Juveniles on coastal intertidal reef flats and mimic similar sized-striped *Halichoeres* wrasses to sneak up on prey. Adults occur on reef crests from coastal to outer reefs. Some variation in colour with coastal forms having a more brownish look, rather then a clean black and white image. Usually solitair but may form schools at times, perhaps for spawning purposes.

Lutjanus semicinctus Quoy & Gaimard, 1824

Half-barred snapper

Length: To 35 cm.
Distribution: East Indonesia, ranging to the Philippines and to Tahiti.
Depth: 10-36 m.
General: Coastal and outer reef crest like the Checkered snapper and seems to replace this species in many areas. A shy species and only comes within photographic range if the diver can sit still and hold its breath long enough, and only if the fish decides.

Lutjanus biguttatus (Valenciennes, 1830)

Two-spot snapper

Length: To 30 cm.
Distribution: Tropical west Pacific and eastern Indian Ocean.
Depth: 3-30 m.
General: Coastal reefs and lagoons. Sometimes forming great schools but large adults often solitair and sometimes in small loose aggregations sheltering with coral plates on outer reef drop-offs. Juveniles are often in sheltered lagoons over sand near seagrasses, have a very pale color but the two back spot remain distinct. A rare xantic form occurs on muddy coastal reefs.

SNAPPERS • LUTJANIDAE

Moluccen snapper

Length: To 30 cm.
Distribution: Widespread tropical west Pacific.
Depth: 5-50 m.
General: A schooling species, sometimes in such large numbers and so dense that entire section of reef are covered. Mainly coastal reefs with little activity during the day, schools often densely packed and floating almost motionless mid-water. Juveniles solitair or small groups in estuaries and near fresh water run-offs.

Lutjanus boutton (Lacépède, 1803)

Sailfin snapper

Length: To 60 cm.
Distribution: Widespread tropical west Pacific.
Depth: 5-60 m.
General: Both adults and juveniles found in similar habitats, sandy flats with coral bommies between reefs or in deep lagoons. Juveniles and large adults which develops an almost vertical head profile shown below.

Symphorichthys spilurus (Günther, 1874)

FUSILIERS • CAESIONIDAE

A small tropical family, comprising 4 genera and 20 species, some ranging to warm-temperate zones, restricted to the Indo-Pacific region. They are closely related to the Lutjanid snappers but are schooling, planktivorous, smallish fishes with streamlined bodies. On the reefs these fishes have ornamental value with their bright colours and movement, as large schools commonly patrolling the slopes and walls in pursuit of zoo-plankton. Differences between genera are mainly in differences of the mouth and the 3 genera included here are difficult to define from external features. Usually *Caesio* is deeper bodied and more robust and *Gymnocasio* most slender. The genus *Pterocaesio* is variable in shape from slender to moderately deep bodied.

Egg and larvae development is identical to the lutjanid snappers with small pelagic eggs, less than 1 mm in diameter and hatchlings are about 2 mm long. Postlarvae are about 30 mm when settling on the substrate, forming small groups on the base of shallow reefs.

Robust fusilier

Length: To 40 cm.
Distribution: Tropical west Pacific and east Indian Ocean, including all of our area.
Depth: 3-40 m.
General: Colour varies slightly except at night or after death, usually becoming bright red. Several similar species, other more slender and lack indistinct stripes below eye, or lack complete yellow tail. Coastal reefs and lagoons, usually in large schools. This is one of the larger species.

Caesio cuning (Bloch, 1791)

Giant fusilier

Length: To 48 cm.
Distribution: Tropical Indian Ocean, from Sri Lanka to Thailand and Java, Indonesia.
Depth: 15-80+ m.
General: Deep rocky reefs, mainly on sea-mounts off shore where they school in large numbers and commonly caught on line by fisherman. Juveniles (smallest seen about 15 cm) identical to adult in colour, swimming fast along reefs in shallower depths in coastal waters. In the southern Sunda Strait the closely related *C. cuning* was occasionally mixed with the schooling adults, which left no doubt about them being different species.

Caesio erythrogaster Cuvier, 1830

FUSILIERS • CAESIONIDAE

Yellow-back fusilier

L: To 30 cm. Di: Indian Ocean, ranging to Java, Indonesia. De: 2-50 m. Ge: Very similar to *C. teres* (below) but yellow colour extends posteriorly to dorsal fin, sometimes over head. Coastal to outer reefs along slopes and drop-offs. Adults common deep off-shore.

Caesio xanthonota Bleeker, 1853

Gold-band fusilier

L: To 35 cm.
Di: Indo-Pacific, including all of our area. De: 3-25 m.
Ge: Coastal reef slopes and inner reefs, along drop-offs forming loose schools, feeding on zooplankton. The closely related Indian Ocean *C. varilineata* ranges to Bali (below).

Caesio caerulaurea Lacépède, 1801

Blue-dash fusilier

Length: To 25 cm.
Distribution: Indo-Pacific, including all of our area.
Depth: 6-25 m.
General: A broad blue band along entire body, very bright with ambient light, and the lack of black caudal fin tips readily identifies this species. At night the lower body half turns bright red. Occurs usually in large schools along deep drop-offs and slopes.

Pterocaesio tile (Cuvier, 1830)

FUSILIERS • CAESIONIDAE

Black-tipped fusilier

L: To 30 cm. Di: Indo-Pacific, including all of our area.
De: 3-25 m. Ge: The 2 thin yellow lines and black tips on caudal fin identifies this species. Below: Indian Ocean species, probably undescribed, taken in southern Java, Indonesia.

Pterocaesio digramma (Bleeker, 1865)

Yellow-band fusilier

Length: To 30 cm.
Distribution: Indo-Pacific, including all of our area.
Depth: 3-25 m.
General: Colour slightly different between juveniles and adults, increasingly broader yellow band with age. Rich coral reef slopes, coastal to outer reef lagoons, usually in small groups. A similar species, *P. lativittata* Carpenter, 1987, occurs in our area, it has yellow above and below most of the lateral line.

Pterocaesio chrysozona (Cuvier, 1830)

Yellow-dash fusilier

Length: To 22 cm.
Distribution: Indonesia, Malaysia and Philippines.
Depth: 6-35 m.
General: The bright yellow dash on the sides readily identifies this species. It remains visible at considerable distance and schools appear as yellow flashes against a dark blue background. Usually near the top of walls along deep outer reefs, in large schools which are nearly always on the move.

Pterocaesio randalli Carpenter, 1987

SILVER BELLIES • GERREIDAE

A small mostly tropical family of silvery fishes, comprising 7 genera and an estimated 40 species, which is most diverse in American waters where they are called mojarras. Generally small to medium sized fishes, inhabiting shallow coastal waters and estuaries, usually schooling oves sandy areas, feeding on bottom invertebrates. The mouth is greatly protrusible, into a downward directed tube, and usually eyes are large, placed immediately behind mouth. They are oval-elongate with moderate sized deciduous cycloid scales, a smallish head and have a long based dorsal fin, which in most genera is anteriorly elevated. Many similar species, usually shiny silver with some dusky markings or black tips on fins, juveniles generally spotted or with faint dark bars.

Eggs are pelagic and tiny, just over o.5 mm in diameter. Postlarvae settle at about 15 mm in shallow coastal, usually brackish habitats.

Deep-bodied silverbelly

Length: To 25 cm.
Distribution: Indonesia, Malaysia to Japan and Micronesia.
Depth: 1-40 m.
General: Protected sand flats, coastal and islands, off beaches or adjacent to reefs. Singly or in small groups, juveniles estuarine and often very shallow, just below intertidal zone. Easily recognised by the deep body and yellow fins below.

Gerres abbreviatus Bleeker, 1850

Slender silverbelly

Length: To 20 cm.
Distribution: Widespread tropical Indo-Pacific, throughout our area.
Depth: 0.2-15 m.
General: Shallow protected coastal and island bays and beaches. Juveniles commonly seen of low tide sand flats between reef and shore, usually in small loose groups. Adults on slightly deeper sand slopes and flats, mainly solitair. Juveniles have a black tip on the dorsal fin and this species is fairly slender.

Gerres oyena (Forsskal, 1775)

SWEETLIPS • HAEMULIDAE

The family comprises the subfamilies Haemulinae, estuarine grunters, and Plectorhinchinae, sweetlips or thicklipped grunters. A large family with an estimated 120 species in about 18 genera. The taxonomy was in a confusing state because of very different phases between juveniles and adults some of which only recently linked. The genus *Diagramma* thought to be monotypic for a long time, several species are now recognised of which at least two occur in our area, both included here.

Sweetlips are mostly reef dwellers, sheltering in caves and often shipwrecks during the day, but feed mainly at night when venturing over sand and rubble in search of benthic invertebrates. They occur singly or in groups, depending on the species or area and are pelagic spawners. Postlarvae settle at a very small size, often less than 10 mm in length. Small juveniles are solitair, have their own unique colour pattern. Most swim close to the substrate in an unusual way by waving the tail in an exaggerated manner, by which they are clearly haemulids. The adults differs from related or similar fish by its small mouth with thick lips. Other features are small scales, strong fin spines and a large head.

Slate sweetlips
L: To 1 m. Di: Indo-Pacific, throughout our area. De: To 30 m. Ge: Juveniles singular, forming small to large groups when loosing the striped pattern. Coastal and inner reefs. Adults usually seen hovering in small groups above reefs or in caves during the day.

Diagramma labiosum MacLeay, 1883

Painted sweetlips
L: To 90 cm. Di: Indonesia to tropical Japan. De: 10-50 m. Ge: Juveniles singular. Adults in small aggregations. Coastal on mud slopes with isolated outcrops of reef or shipwrecks. Differs from the Slate sweetlips by the more spotted pattern and large juveniles are more yellow (below).

Diagramma pictum (Thunberg, 1792)

SWEETLIPS • HAEMULIDAE

Harlequin sweetlips

Length: To 60 cm.
Distribution: Indo-Pacific, throughout our area.
Depth: 3-50 m.
General: Large juveniles and adults may share large ledges and caves along drop-offs. Small juveniles are found in protected bays and lagoons. Dramatic changes from juvenile (below), brown with large white patches, to intermediates with large black spots. The spots reduce in size with age and large adults have small spots over most of the body and head.

Plectorhinchus chaetodonoides Lacépède, 1800

Magpie sweetlips

L: To 90 cm. Di: Indo-Pacific, throughout our area.
De: 10-30 m. Ge: Near outer reefs with clean sand and large coral plates for shelter during the day. Shown a young adults, some juvenile pattern still evident. Small juvenile below.

Plectorhinchus picus (Cuvier, 1930)

SWEETLIPS • HAEMULIDAE

Oriental sweetlips
L: To 50 cm. Di: Indo-Pacific, throughout our area.
De: 3-50 m. Ge: Juveniles solitair among boulders or low coral reefs in shallow depths. Adults form schools on current prone coastal and inner reefs along slopes or the bottom of drop-offs. Juvenile shown below.

Plectorhinchus orientalis (Bloch, 1793)

Lined sweetlips
L: To 40 cm. Di: Indo-Pacific, throughout our area.
De: 1-50 m. Ge: Juveniles solitair, often in shallow seagrass lagoons. Adults form small groups in deep lagoons and in caves along drop-offs on outer reefs. Until recently authors used either *lineatus* of *diagrammus* for this species. Juvenile below.

Plectorhinchus lessoni (Cuvier, 1830)

Oblique-banded sweetlips
L: To 60 cm. Di: Indo-Pacific, throughout our area. De: 3-50 m. Ge: Juveniles solitair in lagoons or coastal on boulder or low coral reef slopes. Adults more on outer reefs along drop-offs and in some areas form large schools. Until recently the adult form was known as *goldmanni*. Below a 20 cm juvenile.

Plectorhinchus lineatus (Linnaeus, 1754)

SWEETLIPS • HAEMULIDAE

Yellow-ribbon sweetlips

L: To 50 cm. Di: Widespread in our area. De: 3-50 m. Ge: Juveniles solitair on deep coastal rubble slopes. Adults form small to large groups or mix with similar species. Number of stripes increase with age. Coastal, current-prone reefs and preferring the deeper part of its range. Juvenile shown below.

Plectorhinchus polytaenia (Bleeker, 1852)

Orange-lined sweetlips

Length: To 40 cm.
Distribution: West Pacific, throughout our area.
Depth: 3-20 m.
General: Mainly in coastal reefs and lagoons. Usually in small groups swimming above reefs during the day in small aggregations, but forms schools in some areas. Juveniles are similar but have fewer and thicker lines.

Plectorhinchus celebicus Bleeker, 1873

Gold-spotted sweetlips

Length: To 60 cm.
Distribution: Indo-Pacific, throughout our area.
Depth: 3-30 m.
General: Coastal reefs and bays, often in silty areas. Juveniles in seagrass beds and have longitudinal bands, similar to a *Lutjanus* snapper.

Plectorhinchus flavomaculatus (Cuvier, 1830)

SWEETLIPS • HAEMULIDAE

Plectorhinchus obscurus (Günther, 1871)

Giant sweetlips

Length: To 100 cm.
Distribution: Indo-Pacific, throughout our area.
Depth: 20-60 m.
General: Deep coastal reefs, usually along the base of drop-offs and on off-shore submerged reefs. The largest sweetlips, often in small groups.

Plectorhinchus gibbosus (Lacèpède, 1802)

Brown sweetlips

Length: To 60 cm.
Distribution: Indo-Pacific, throughout our area.
Depth: 1-50 m.
General: Coastal, often silty habitat and not often seen in clear waters as in the photograph. Small juveniles float with debris or weeds on the surface.

Plectorhinchus sp

Andaman sweetlips

This unsual sweetlips was photographed in the Andaman Sea and appears to be a new species. It mixes with *P. orientalis* and shows similarity in colour to *P. chaetodonoides* but hybrids are unlikely.

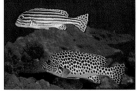

EMPERORS • LETHRINIDAE

The Emperors and Large-eye breams are medium sized fishes, comprising 5 genera and 39 species, distributed over the Indo-Pacific, except one that only occurs in the eastern Pacific. The largest genus is *Lethrinus* and many of which encountered in shallow depths. These fishes are mainly found on or near tropical reefs and only a few range into warm-temperate waters.

The *Lethrinus* species are also known as Emperor-sweetlips or Emperor-snapper, whilst the remaining genera are mainly known as Large-eye breams. They are bottom dwellers, carnivorous, feeding on a variety of invertebrates and small fishes. Some species feed only at night, resting on reefs during the day, but others feed during both periods and may alternate between open sand and reef. Most species however feed on or near reefs, either solitair or schooling. The Emperors are regarded as first-rate foodfish, but some of the Large-eyed breams, generally excellent to eat, can be effected by iodine which causes cooking problems. In addition in some areas these fishes can be effected by ciguatoxic poisoning.

Eggs are pelagic, small, less than 1 mm in diameter and produced in vast numbers. Larvae hatch at less than 2 mm long and post larvae are about 30 mm long. Many species settle in seagrass beds and are camouflage, green and can change colour quickly. Most juveniles look very different from adults and some have different colour phases in between.

Gold-spot emperor

Length: To 30 cm.
Distribution: Indo-Pacific, containing all of our area.
Depth: 5-30 m.
General: Colour pale in white sand areas and very dark, lacking stripes, along deep drop-offs. Yellow blotch below soft dorsal fin usually remains prominent in all colour variations. A schooling species, sometimes forming aggregations with hundreds of individuals. It feeds mainly on the bottom in reef-lagoons to deep drop-offs. Also known as Gold-Lined Sea Bream or Glowfish.

Gnathodentex aurolineatus (Lacépède, 1802)

EMPERORS • LETHRINIDAE

Slender emperor

L: To 20 cm. Di: Indo-Pacific, throughout our area. De: 1-30 m. Ge: Coastal algae reef and young in seagrasses. More slender than most other species. The most similar Lancer *L. genivittatus* (below) is also coastal but sand-reef.

Lethrinus variegatus Valenciennes, 1830

Small-tooth emperor

L: To 70 cm. Di: Widespread tropical Indo-Pacific, containing all of our area. De: 3-80 m. Ge: Coastal sand slopes with low reef patches, adjacent to deep water. Changes colour quickly to a mottly pattern when stopping to feed, as shown below by the almost identical but longer snouted *L. olivaceus*.

Lethrinus microdon Valenciennes, 1830

Yellow-striped emperor

Length: To 40 cm.
Distribution: Widespread tropical Indo-Pacific, containing all of our area.
Depth: 1-30 m.
General: A shy species, singly or loosely in small numbers, mainly on coastal rocky reef flats and slopes, feeding on small sand patches between rocks or corals.
Feeds primarily on benthic invertebrates. Small juveniles near freshwater run-offs where algaes dominate on boulder rock. Easily identified by the yellow stripes.

Lethrinus ornatus Valenciennes, 1830

EMPERORS • LETHRINIDAE

Blue-spotted large-eye bream

Length: To 40 cm.
Distribution: Tropical Indo-Pacific, from Andaman Sea through most of Indonesia to Marshall Islands, and north to Japan. Depth: 20-50 m.
General: Several very similar species in our area, usually silvery, this species has small blue spots on snout and occurs shallower than the others. Coastal sand slopes, feeding on benthic invertebrates along reef margins. Swims well above the bottom, usually a short distance and than stopping motionless to study the bottom for food, diving quickly at prey.

Gymnocranius microdon (Bleeker, 1851)

Big-eye emperor

Length: To 60 cm.
Distribution: Widespread tropical Indo-Pacific, containing all of our area. Depth: 1-60 m.
General: Small juvenile (below) solitair on rubble sand patches in shallow lagoons or on reef crests from clear coastal to outer reef habitats. Adults congregate during the day, hoovering near drop-offs or near large caves, often in large numbers, moving out at night to feed on the substrate in deep water.

Monotaxis grandoculis (Forsskal, 1775)

THREADFIN BREAMS • NEMIPTERIDAE

Two of the five genera in this family have primarily deep water species. *Nemipterus* the larges genus with 26 species and *Parascolopsis* with 9 species are rarely seen by diving and usually in excess of 50 m. depth. *Scolopsis* is mostly encountered with 17 described and one new species, and *Pentapodus* with 11 species, and *Scaevius* is monotypic, the family totalling 65 species. The *Scolopsis* are known as monocle breams, spinecheeks or coral breams, *Pentapodus* as whiptails and species of *Nemipterus* are the threadfin breams.

The species included here are benthic feeders mainly active during the day. At night they rest on the substrate against rocks and corals. They feed on small crustaceans, worms and some on small fishes, which are located with their excellent eye sight, and take mouths full of sand to sift out prey. Most species occur in small but loose aggregations and when feeding swimming just above the substrate with sudden stops to study the surrounding. Juveniles are usually with distinct longitudinal stripes and adults change to a completely different pattern.

Striped whiptail

Length: To 20 cm.
Distribution: Indonesia, Philippines and New Guinea.
Depth: 3-15 m.
General: A solitary species usually seen roaming reef tops, mainly in coastal and inner reef zones, but also enters lagoons and harbours. Juveniles with darker and more distinct stripes, usually found in loose group in very shallow coastal sand flats.

Pentapodus trivittatus (Bloch, 1791)

Butterfly whiptail

L: To 20 cm. Di: Throughout Indonesia, ranging to Philippines, Singapore and south China seas. De: 10-100 m. Ge: Coastal sandy areas between reefs, usually in depths of 15 m or more. The similar widespread occurring *P. caninus* below.

Pentapodus setosus (Valenciennes, 1830)

THREADFIN BREAMS • NEMIPTERIDAE

Bald-spot spinecheek

L: To 30 cm. Di: Sulawesi, Flores and north coast on New Guinea to Fiji. De: 10-30 m. Ge: Rare, occasionally in clean sand areas near reefs but probably prefers large open flats. Often confused with the common **S. monogramma** below.

Scolopsis temporalis (Cuvier, 1830)

Three-line spinecheek

Length: 20 cm.
Distribution: Indo-Pacific, throughout our area.
Depth: 1-20 m. General: Mainly a coastal species found solitair on reef flats and in lagoons on sand near reef. Often near mangrove areas.
The similar **S. lineata** below.

Scolopsis trilineata Kner, 1868

Blue-stripe spinecheek

Length: To 20 cm.
Distribution: Indo-Pacific, throughout our area.
Depth: 5-50 m.
General: Adults in small aggregation on the deeper parts of coastal rubble slopes, swimming near the bottom in search of small animal prey, mainly crustaceans. Juveniles solitair and in similar habitat as adults but usually in shallower depths. They are easily distinguished from other striped juveniles by the blue and black lines along their upper sides.

Scolopsis xenochroa Günther, 1872

THREADFIN BREAMS • NEMIPTERIDAE

Scolopsis affinis Peters, 1877

Yellow-tail spinecheek

Length: To 25 cm.
Distribution: Widespread tropical west Pacific and east Indian Ocean.
Depth: 20-60 m.
General: Coastal and inner reef zones. Usually in small to large aggregations over sand flats near reefs. This, the following two species are often confused or synonymised. Where species co-occur they often mix in schools. Juveniles have a black mid-lateral stripe with borthering yellow above, and adults occasionally show such a stripe.

Scolopsis aurata (Park, 1797)

Golden spinecheek
L: To 25 cm. Di: Indian Ocean, mainly Sumatra and Java.
De: 10-50 m. Ge: Juveniles coastal on sandy flats between reefs. Adults, usually in loose groups on sand and rubble slopes in depths of about 20 m or more. A similar species (below) appears to be undescribed.

Scolopsis ciliata (Lacépède, 1802)

Silver-line spinecheek

Length: To 20 cm.
Distribution: Tropical west Pacific.
Depth: 1-50 m.
General: Mainly in small groups on shallow to deep coastal reef slopes, hunting small benthic animals near the substrate. The silver line near the dorsal fin and yellow eye easily identifies the adult of this species. Juveniles are similar to adults but have an additional white mid-lateral stripe.

GOATFISHES • MULLIDAE

An easily identified family by the presents of a pair of barbels on the chin, comprising six genera and about 35 species with a global distribution in tropical and sub-tropical seas. They typically have elongated bodies with large scales, all fins are pointed or angular, two separate dorsal fins and a forked tail. Except for colour many species of the same genus are difficult to separate on morphological features alone.

At night most species change colour to such extend that they are difficult to identify. Many are bright red all over or with broad bands. Sometimes these colours show during the day when resting on the substrate or when visiting cleaning stations to highlight the presence of parasites. Their daytime colour quickly returns when swimming of. Some species change even from swimming along with a general plain look to a feeding colour on the substrate with turning on a blotched or barred pattern for camouflage.

Goatfishes are benthic feeders and use their strong barbels to dig and sense prey in the substrate. The larger species dig by themselves or have a few companions, but often these consist of unrelated species such as wrasses and even carangids which keep on eye on the works in case something tasty appears. Some of the smaller species and juveniles of the larger species school and may travel along substrated whilst feeding at a rapid rate. On mud this can created a large turbid colomn of water behind them. They are pelagic spawners and produce tiny eggs, less than 1 mm diameter. Pelagic larvae can get quite large and reach 50 mm before settling down.

Yellow-saddle goatfish
L: To 40 cm. Di: Indo-Pacific, throughout our area.
De: 3-100 m. Ge: Various habitats from shallow coastal to deep off shore. Adults singly or in pairs, sometimes in small groups. Juveniles solitair but usually mixes with other species such as wrasses, especially of the genus *Thalassoma*. Juveniles maybe bright yellow and this form is sometimes retained.

Parupeneus cyclostomus (Lacépède, 1801)

GOATFISHES • MULLIDAE

Parupeneus heptacanthus (Lacépède, 1801)

Small-spot goatfish

Length: 35 cm.
Distribution: Indo-Pacific, throughout our area.
Depth: 3-60 m.
General: A common coastal species, usually in muddy or silty habitats. Juveniles are similar to adults but more slender, and form small aggregations on shallow flats and adults usually deeper along slopes, either singly or in small groups. May appear to be almost white in colour with the ambient light, especially in depths below 10 m.

Parupeneus indicus (Shaw, 1803)

Yellow-spot goatfish

Length: To 40 cm.
Distribution: Indo-Pacific, throughout our area.
Depth: 5-25 m.
General: Occurs in coastal waters and inner reef lagoons, usually in small agregations of less than ten individuals, sometimes solitair. Juveniles are similar to adults but more slender. The yellow blotch on the side identifies this species.

Parupeneus barberinus (Lacépède, 1801)

Dash and dot goatfish

Length: To 40 cm.
Distribution: Indo-Pacific, throughout our area.
Depth: 5-100 m.
General: Coastal reefs and lagoons, solitair or in small numbers, and often followed by wrasses. Reports of this species growing larger than here indicated, appears to be based on exaggeration. Juveniles whitish with just a black stripe. Adults with or without yellow above the stripe.

GOATFISHES • MULLIDAE

Long-barbel goatfish

Length: To 30 cm.
Distribution: Indo-Pacific, throughout our area.
Depth: 1-25 m.
General: A coastal reef species, preferring rocky areas, and often found solitair or in pairs. Like most species in the genus the colour of the body can turn bright red and this species seems to do this more often than others.
It is easily overlooked as it lookes very much like the more common *P. barberinus* but the stripe is thicker and followed by a pale area which is uniform with the rest of the body in the former.

Parupeneus macronema (Lacépède, 1801)

Banded goatfish

Length: To 35 cm.
Distribution: Indo-Pacific, throughout our area.
Depth: 1-140 m.
General: Observed almost everywhere on sand and rubble near reefs at all depths divable, singly or in small groups. Juvenile similar but as typically to the genus, more slender. Best identified by the black areas below second dorsal fin and on caudal peduncle.

Parupeneus multifasciatus (Quoy & Gaimard, 1824)

Double-bar goatfish

Length: To 30 cm.
Distribution: Indo-Pacific, throughout our area.
Depth: 1-80 m.
General: A coral reef dweller, usually solitair, often seen resting on corals. Coastal reef tops and usually seen in depths between 3-10 m. Some geographical variations, juveniles with two distinct bars but in adults these can be reduced to short saddles or indistinct barring.

Parupeneus bifasciatus (Lacépède, 1801)

GOATFISHES • MULLIDAE

Parupeneus pleurostigma (Bennett, 1830)

Round-spot goatfish

Length: To 30 cm.
Distribution: Indo-Pacific, throughout our area.
Depth: 10-40 m.
General: Clear water reefs in lagoons and on slopes. Usually singly or in small groups. juveniles coastal and often in small aggregations, and similar in colour to adult but typically much more slender. The large black spot directly followed by the white area should readily identify this species as adult.

Parupeneus barberinoides (Bleeker, 1852)

Half & half goatfish

Length: To 30 cm, but rarely over 20 cm.
Distribution: Indo-Pacific, throughout our area.
Depth: 3-40 m.
General: The smallest species in the genus and with the most unusual colouration which makes it easy to identify. A solitary species which as juvenile occurs in shallow coastal bays in algae habitats and adults move into deeper water, usually found on rubble slopes with mixed coral and algae growth.

Upeneus tragula Richardson, 1846

Bar-tail goatfish

Length: To 30 cm.
Distribution: Indo-Pacific, throughout our area.
Depth: 3-30 m.
General: Coastal reef slopes and lagoons. Adults in small groups or solitair. Several similar species, some undescribed and most of the smaller ones occur in large schools in muddy and coastal areas, and in addition there are some which prefer deep water and are only known from trawls. This species can change colour rapidly and when swimming it turns plain white with a dark mid-lateral stripe.

GOATFISHES • MULLIDAE

Striped goatfish

Length: To 25 cm.
Distribution: Indo-Pacific, throughout our area.
Depth: 5-100 m.
General: A very common schooling species in muddy coastal areas on sloped and deep flats. One of several similar species which feed by numerous individuals side by side forming a long line, moving along at a rapid pace and usually creating cloudy water conditions in the process.

Upeneus vittatus (Forsskal, 1775)

Yellow-stripe goatfish

Length: To 35 cm.
Distribution: Indo-Pacific, throughout our area.
Depth: 5-100 m.
General: A schooling species, usually in moderate to large numbers hoovering or swimming slowly over reefs during the day, to move to sand flats at night to feed. Often mixed in schools with yellow lutjanid species such as *kasmira*. Juveniles sometimes solitair but soon form small groups. *Mulloides* used by many authors is a synonym of *Arripis*.

Mulloidichthys vanicolensis (Valenciennes, 1831)

Square-spot goatfish

Length: To 40 cm.
Distribution: Indo-Pacific, throughout our area.
Depth: Intertidal to 30 m.
General: Juveniles in small groups in shallow coastal waters, including rock pools. Adults on deeper sand or mud slopes in coastal areas and often solitair.
This species is distinct by the small square black spot.
A yellow lateral line passing through the spot is often very distinct, especially in adults.

Muloidichthys flavolineatus (Lacépède, 1801)

BULLSEYES • PEMPHERIDIDAE

Two genera are currently recognised and both are represented here. An estimated 20 species are distributed globally in tropical to warm-temperate seas. They are closely related to the silver batfishes, Monodactylidae, and have many features in common. Bodies are oblong to moderately slender and compressed. In the genus *Pempheris* the lateral-line scales extend to posterior margin of caudal fin. Small to large schools aggregate near reefs or in large caves during the day and during the night feed well away in open water on zoo-plankton, usually small crustaceans and cephalopods. Small juveniles are semi-transparent, often in very large cloud-like formations along the front of small reefs in coastal estuaries. Adults are usually secretive, staying in the shelter of the reef or live deep inside.

Eggs are presumed pelagic and postlarvae settle at a very small size, only about 10 mm long, forming schools in sheltered reefs in coastal bays and open estuaries. Also called sweepers.

Pempheris vanicolensis Cuvier, 1831

Greenback bullseye

Length: To 12 cm.
Distribution: Tropical Indo-Pacific, throughout our area.
Depth: 3-20 m.
General: Secretive during the day, in large numbers amongst large rocky boulders in shallow coastal areas, or in large caves. Best identified by colour and straight lateral-line with short curve at origin.

Pempheris adusta Bleeker, 1877

Black-base bullseye

Length: To 17 cm.
Distribution: Tropical Indo-Pacific, throughout our area.
Depth: 10-30 m.
General: Mostly found in large schools in the back of spacious overhangs along drop-offs. Two very similar species, this one in which the lateral line has a large curve from origin rising above straight section and the other in which the curve evens out at straight level, *P. schwenkii* Bleeker, both in similar habitats.

BULLSEYES • PEMPHERIDIDAE

Black-margin bullseye

Length: To 16 cm.
Distribution: Tropical Indo-Pacific, throughout our area.
Depth: 3-20 m.
General: During the day found along drop-offs and slopes in small caves or under overhangs and often solitair, clear coastal to outer reefs. This species has a straight lateral line but long curve at origin.
Very similar to *P. oualensis* which has a black spot on pectoral fin base and is found in the same habitats in our area.

Pempheris mangula Cuvier, 1829

Yellow sweeper

Length: To 10 cm.
Distribution: Most of Indonesia, fading towards Indian Ocean to west Pacific, throughout the Pacific part of our area.
Depth: 10-50 m.
General: A nocturnal species, forming large dense schools, often filling entire large caves along drop-offs, or deep around isolated bommies on sand remote from main reefs. Disperse at night to feed in open water on zooplankton.

Parapriacanthus ransonneti (Steindachner)

DRUMMERS • KYPHOSIDAE

A small family, comprising 3 genera, 2 of which monotypic are not included here. The genus *Kyphosus* comprises an estimated 8 species, which have a global distribution. Kyphosids are primarily tropical reef fishes, commonly found in high energy zones in shallow coastal waters where they feed on algae and associated invertebrates. Mostly in large boulder areas where they go for cover and usually these medium sized fishes occur in large schools. Occasionally they are caught by line fisherman but are considered to be of poor taste. Reputed in some areas to cause vivid nightmares after consumption. Some species have a xanthic form.

All species congregate in great numbers to spawn and produce vast numbers of tiny eggs. Postlarvae settle in estuaries in weedy areas but may stay pelagic with floating weeds for a long time before reaching shores. These fishes are known under a variety of names in different areas because of their great distribution. Sea chubs is used in the Americas, and Rudderfishes in some other areas of the Indian Ocean.

Brassy drummer

L: To 40 cm. Di: Indo-Pacific, throughout our area.
De: Surface to 20 m.
Ge: Coast reef flats, slopes and lagoons. Swims mid water, often singly or in small aggregations. Very shallow, to outer reefs in exposed areas.

Kyphosus vaigiensis (Quoy and Gaimard, 1824)

Snubnose drummer

L: To 50 cm. Di: Indo-Pacific, throughout our area.
De: Surface to 20 m.
Ge: Usually in large schools, often swimming mid water, but in small numbers when resident on some reefs. Shallow coastal reefs and lagoons, feeding mainly on free floating algae.

Opposit page:
Tiny juvenile tall-fin batfish *Platax teira* on the surface, typical for several species to mimic leaves from trees in coastal waters. Flores, Indonesia.

Kyphosus cinerascens (Forsskal, 1775)

BATFISHES • EPHIPPIDAE

A small but interesting family comprising several genera, of which only *Platax* is commonly seen in our area. Others are found deep or in the Atlantic. All known five species of *Platax* are included here. Only recently a fifth species, *P. boersi*, was recognised as being distict, it was confused with other species.

Adult fish can be found from coastal lagoons to outer reefs where they may form large schools, and have been trawled to depths of 500. They are thought to be pelagic spawners and larvae transforms to the fimiliar, sometimes very tall, juvenile when about 20 mm long. Juveniles are either pelagic or benthic, depending on the species, usually seeking quiet waters in coastal bays and estuaries. Diet consist of algae and a variety of invertebrates such as jellies or other plankton.

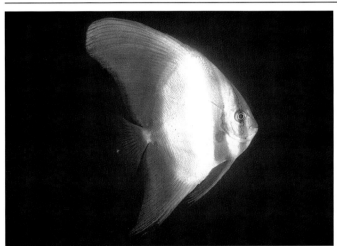

Round batfish

Length: To 50 cm.
Distribution: Widespread tropical Indo-Pacific, throughout our area.
Depth: Surface to 40 m.
General: Small juveniles float against the surface with current or wind-driven land debris, in particularily orange-brown leaves. They settle in lagoons or along beaches on sand in small schools. Adults pair or school in different areas and usually found in deeper off shore areas.
Small juvenile below.

Platax orbicularis (Forsskal, 1775)

BATFISHES • EPHIPPIDAE

Shaded batfish
L: To 35 cm. Di: Tropical west Pacific, throughout our area. De: 2-50 m. Ge: Post larvae settle on coastal protected reefs or in mangroves. Juveniles are often under jetties but also move to deep water where they inhabit caves. Adults are sometimes solitair, in small groups or in large schools from shallow coastal to deep outer reefs, and often seen in shipwrecks. Juvenile below.

Platax pinnatus (Linnaeus, 1758)

Zebra batfish
L: To 65 cm. Di: Tropical west Pacific, throughout our area. De: 15-60 m. Ge: A large species, usually solitair or in small loose groups in deep lagoons or sandflats between reefs. Juveniles with crinoids and the individual in the photograph was observed nibbling on the arms. Adult known as humped batfish are either solitair or in small groups. The least observed batfish species. Juvenile below.

Platax batavianus Cuvier, 1831

BATFISHES • EPHIPPIDAE

Tall-fin batfish
L: To 60 cm. Di: Very widespread in the tropical and subtropical Indo-Pacific, containing all of our area. De: Surface to very deep. Ge: Small juveniles under floating subjects, forming schools as they find each other. Adults solitair or in schools often with shipwrecks, quickly adapting to visiting divers and often rise to the surface to meet them when starting to descend. Juvenile below.

Platax teira (Forsskal, 1775)

Boers batfish
L: To 40 cm. Di: Indonesia, New Guinea and Philippines. De:1-50 m. Ge: Adults usually in large schools along dropoffs, including outer reefs, but also observed singly in coastal waters. Because of the shape usually confused with the round batfish, *P. orbicularis*. Small juveniles easily confused with Tall-fin batfish *P. teira* and best separated by the shape of the median fins. Juvenile below.

Platax boersii Bleeker, 1852

SILVER BATFISHES • MONODACTYLIDAE

Silver batfish

Length: To 20 cm.
Distribution: Indo-Pacific, including all of our areas.
Depth: Surface to 10 m.
General: Colour dark in juveniles and silver in adults. Occurs mainly in large coastal estuaries in silty habitats, often in large schools around break waters or under jetties. Juveniles in brackish water, sometimes moving into freshwaters. Adults move about from river mouths to coastal reefs.
A small family with 3 genera and 5 species, only *Monodactylus* in our area. They have tiny ventral fins which become rudimentory in adults.

Monodactylus argenteus (Linnaeus, 1758)

SCATS • SCATOPHAGIDAE

Spotted scat

Length: 35 cm.
Distribution: Indo-Pacific, throughout our area.
Depth: Surface to 10 m.
General: Variable stages with age. Small juveniles black with red, becoming spotted, first forming bands, and decreasing to small spots in adults. Coastal estuaries and river entrances, mainly in brackish water, including settling juveniles. Also known as Tiger Scat and Spotted Butterfish. Small family, 2 genera and 3 or 4 species. Compressed bodies, similar to butterflyfishes, but 4 spines in anal fin. The fin spines can inflict painful wounds and are thought to be venomous.

Scatophagus argus (Linnaeus, 1766)

BUTTERFLYFISHES • CHAETODONTIDAE

This large family of colourful fishes is popular with divers and aquarists. It comprises 10 genera with about 120 species which mostly inhabit coral reefs, but a few have adapted to warm-temperate zones, and some live in deep water. The majority of species occur in the tropical Indo-Pacific with the largest concentration on Indonesian coral reefs, only 4 are known from the east Pacific and 12 from Atlantic seas. Most species are placed *Chaetodontidae* with 114 species in 13 subgenera, *Heniochus* has 8 species and the remainder are monotypic or have a few species each. Common butterflyfishes known from the Western Indian Ocean are not included.

Most species occur in shallow depths on coral reefs, many of which in pairs or in schools. Some species pair in one area and school in an other. Their diet consists of small invertebrates which are usually picked from the substrate, except those adapted to zooplankton. They produce tiny spherical pelagic eggs less than 1 mm in diameter and larvae have a bony head-armour, often with serrated spines, which is known as the tholichthys stage. Post-larvae can be as small as 10 mm long and settle in corals and rocks and most species are very secretive.

Black-barred butterflyfish

Length: To 14 cm.
Distribution: Indonesia (Flores, Sulawesi) to Palau region.
Depth: 20-75 m.
General: Lives on outer reef drop-offs, usually in depths in excess of 40 m, but the specimen in photograph was taken in only 20 m depth in Sulawesi. In this particular area, a drop-off which featured very large sponges growing outwards from the drop-offs, it was common, occurring only in pairs.

Chaetodon burgessi Allen & Starck, 1973

Cross-hatch butterflyfish

Length: To 10 cm.
Distribution: Indonesia and Philippines.
Depth: 6-50 m.
General: Adults pair along deep slopes and drop-offs, mainly on outer reefs. Actively moves along over large areas to various depths in search for food, stopping and inspecting particular areas on the way. Juveniles solitair, secretive in ledges along drop-offs and remaining in small areas.

Opposit page:
Head-band butterflyfish
Chaetodon collare schooling in Andaman Sea.

Chaetodon xanthurus Bleeker, 1857

BUTTERFLYFISHES • CHAETODONTIDAE

Chaetodon melannotus Bloch & Schneider, 1801

Black-back butterflyfish

Length: To 15 cm.
Distribution: Widespread tropical Indo-Pacific, including all of our area.
Depth: 1-30 m.
General: A pairing species but occasionally groups of 10 or more individuals can be seen traveling over reefs to new territories. Juveniles secretive on shallow coastal, usually rocky reefs. Mostly on coastal reefs on slopes near drop-offs.

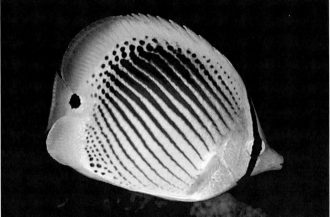

Chaetodon ocellicaudus Cuvier, 1831

Tail-spot butterflyfish

Length: To 14 cm.
Distribution: Indonesia, Philippines to Andaman Sea.
Depth: 5-50 m.
General: Solitair or in pairs, usually on steep slopes and drop-offs, sometimes entering cuttings of reef crests. Coastal to outer reefs. The distinct black spot on the caudal peduncle separates it from the similar *C. melannotus* above, which in the juvenile stage is even more pronounced.

Chaetodon citrinellus Cuvier, 1831

Citron butterflyfish

Length: 11 cm.
Distribution: Widespread tropical Indo-Pacific, including all of our area.
Depth: 1-36 m.
General: Occurs in pairs, but regularily school, seemingly when on the move to other areas. Juveniles secretive and often seek safety by hiding behind sea-urchins. Seems to prefer crear water reefs but can be found in coastal as well as outer reefs and in various depths.

BUTTERFLYFISHES • CHAETODONTIDAE

Spot-banded butterflyfish

Length: To 10 cm.
Distribution: Widespread Indonesian waters and Philippines.
Depth: 6-45 m.
General: Usually seen in pairs along steep drop-offs on outer reefs. An active and often shy species, quickly moving along and feeding on small invertebrates picked off coral-rock. Juveniles very secretive in narrow ledges as shown in the lower photo from the Talaud-Islands, Northern Indonesia. This species has an unusual coloration.

Chaetodon punctatofasciatus Cuvier, 1831

Eye-patch butterflyfish

Length: To 20 cm.
Distribution: West Pacific, ranging from north-west Australia and Java to southern Japan.
Depth: 3-30 m.
General: Coastal reef tops and slopes in hydroid and soft coral areas. Usually in small groups or pairs, small groups often swimming midwater showing little or no activity during the day. Juveniles similar to adults and solitair on reefs.

Chaetodon adiergastos Seale, 1910

BUTTERFLYFISHES • CHAETODONTIDAE

Reticulated butterflyfish
L: To 20 cm. Di: Northern Indonesia to Philippines, northern New Guinea to New Caledonia. De: 1-30 m. Ge: An oceanic species with a preference to clear water reefs. The similar Indian ***C. collare*** (below) is common to Bali the most eastern part of its range.

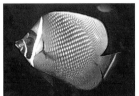

Chaetodon reticulatus Cuvier, 1831

Raccoon butterflyfish

Length: To 20 cm.
Distribution: Widespread tropical Indo-Pacific, including all of our area.
Depth: 1-25 m.
General: Juveniles on coastal reefs in in harbours, secretivly in shallow areas, usually 1-2 m depth. Adults in various habitats from coastal to outer reefs and commonly found on shipwrecks. They pair in Indonesia, but school in some oceanic localities.

Chaetodon lunula (Lacépède, 1803)

Ornate butterflyfish

Length: To 20 cm.
Distribution: Tropical west Pacific + Andaman Sea.
Depth: 1-40 m.
General: Almost always in pairs swimming over reef crests near drop-offs. Feeds on corals, nibbling the polyps. The similar ***C. meyeri*** is shown below.

Chaetodon ornatissimus Cuvier, 1831

BUTTERFLYFISHES • CHAETODONTIDAE

Pacific pinstriped butterflyfish
L: To 12 cm. Di: West Pacific, from Java east throughout our area. De: 3-20 m. Ge: In pairs or small groups in coastal coral rich areas. Small juv. hide amongst branches of small coral heads, mostly in lagoons. The Indian pinstriped butterflyfish **C. trifasciatus** (below) east to Bali.

Chaetodon lunulatus Quoy & Gaimard, 1824

Chevroned butterflyfish

Length: 14 cm.
Distribution: Widespread tropical Indo-Pacific, including all of our area.
Depth: 2-20 m.
General: A unique species place in its own subgenus of *Megaprotodon*. It occurs primarily with *Acropora* plate shaped or table top corals, singly or in pairs and often juveniles are present as well. Juveniles differ in having the black on the tail forward on the body and fins instead. Coastal to outer reefs.

Chaetodon trifascialis Quoy & Gaimard, 1824

Blue-dash butterflyfish
L: To 12 cm. Di: From Australia through our range to southern Japan, Andaman Sea? De: 2-15 m. Ge: Singly or in pairs clear waters, feeding on coral polyps. Juveniles secretively between coral branches. Below (photo Andaman Sea) the Indian Ocean form which may represent a new species.

Chaetodon plebeius Cuvier, 1831

BUTTERFLYFISHES • CHAETODONTIDAE

Oval-spot butterflyfish
L: To 16 cm. Di: West Pacific to Indian Ocean, all of our area. De: 3-20 m. Gen: Coastal, occurring in pairs or small aggregations, feeding primarily on soft corals. Tiny juveniles among branching corals and often several together in one coral head. The similar Eclipse butterfly *C. bennetti* below.

Chaetodon speculum Cuvier, 1831

Teardrop butterflyfish

L: To 20 cm.
Di: West Pacific to Indian Ocean, throughout our area. Replaced by yellow subspecies further west, below. De: 1-30 m. Ge: In our area nearly always in pairs, uncommon in central Indonesian waters and more abundant oceanic.

Chaetodon unimaculatus Bloch, 1787

Pig-face butterflyfish

L: To 30 cm. Di: West Pacific, throughout Indonesia, north to Philippines and Japan. De: 10-40 m. Ge: Coastal and inner reef slopes with rich coral growth, usually in pairs. The similar *C. lineolatus*, the largest of the butterflyfishes, also in our area below.

Chaetodon oxycephalus Bleeker, 1853

BUTTERFLYFISHES • CHAETODONTIDAE

Pacific double-saddle butterflyfish

L: To 15 cm. Di: West Pacific, ranging west to Java. De: 1-30 m. Ge: In our area usually in pairs or small groups, but schools in some Pacific oceanic locations. Coastal slopes and along drop-offs. Replaced by the sibling Indian Ocean *C. falcula,* below, from Java, Indonesia.

Chaetodon ulietensis Cuvier, 1831

Vagabond butterflyfish

L: To 18 cm. Di: Philippines, Malaysia, Thailand. De: 1-30 m. Ge: Coastal to inner reefs in lagoons and rich slopes. Usually shallow, swimming in pairs close to the substrate grazing on small invertebrates and algae. The sibling **C. *decussatus,*** below.

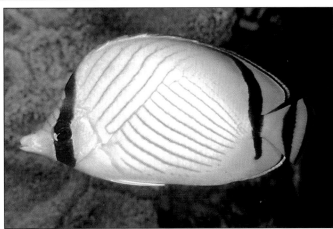

Chaetodon vagabundus Linnaeus, 1758

Pacific triangular butterflyfish

L: To 15 cm. Di: West Pacific, ranging to north Java, Indonesia. De: 3-15 m. Ge: Adults singly or in pairs on reef crests and slopes with *Acropora* corals on which they feed as a major part of their diet. It's Indian sibling *C. triangulum* ranges to Java.

Chaetodon baronessa Cuvier, 1831

BUTTERFLYFISHES • CHAETODONTIDAE

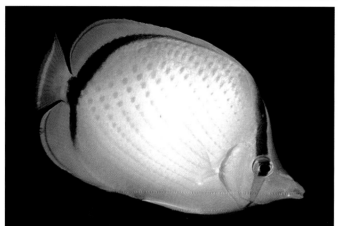

Chaetodon selene Bleeker, 1853

Yellow-dotted butterflyfish

Length: To 20 cm.
Distribution: Indonesia, Malaysia, Philippines to southern Japan.
Depth: 15-50 m.
General: Deep coastal rubble and sand slopes with mixed reef patches. Usually in pairs swimming close to the substrate, feeding on bentic invertebrates. Juveniles solitair with small reef outcrops in depths of about 15 m.

Chaetodon ephippium Cuvier, 1831

Saddled butterflyfish

Length: To 23 cm.
Distribution: Widespread tropical west Pacific and east Indian Ocean, including all of our area.
Depth: 1-30 m.
General: Coastal to outer reefs slopes and crests with rich coral growth. Singly or in pairs and sometimes in small groups which may consist of mixed butterflyfish species. Mostly shallow in about 10 m depth, but in some areas observed deeper. Juveniles in coastal bays and harbours, secretive among rocks or rubble.

Chaetodon rafflesi Bennett, 1830

Lattice butterflyfish

Length: To 20 cm.
Distribution: Andaman Sea to Philippines.
Depth: 1-15 m.
General: Lagoons and reef crest from coastal to outer reefs. Adults usually pair, feeding on the substrate on invertebrates, including worms and corals. Small juveniles secretive in amongst dense branching coral heads.

BUTTERFLYFISHES • CHAETODONTIDAE

Dotted butterflyfish

Length: To 24 cm.
Distribution: Tropical west Pacific including our range.
Depth: 3-50 m.
General: A shy species, usually in pairs on rich and clear coastal reef crests in gutters, channels, along slopes and deep drop-offs feeding on invertebrates from the substrate. Occasionally schools for migrating purposes.

Chaetodon semeion Bleeker, 1855

Orange-banded coralfish

Length: To 14 cm.
Distribution: Widespread tropical west Pacific, including all of our area.
Depth: 5-60 m.
General: Occurs singly or in pairs on deep coastal in inner reef slopes. Seems to prefer cooler water temperatures in equatorial waters and often found in cool up welling areas such as in the straits between Indonesian islands from the Indian Ocean. Juveniles similar but ocellus in dorsal fin much larger. Usually in sponge areas, taking small invertebrates of their surface.

Coradion chrysozonus (Cuvier, 1831)

Two-eyed coralfish

L: To 12 cm. Di: Malaysia, Indonesia and Philippines.
De: 5-50 m. Ge: Usually in pairs. Coastal to outer reefs in the vicinity of barrel sponges It is readily identified by the ocelli on the dorsal and anal fins. The high fin **C. altivelis** below.

Coradion melanopus (Cuvier, 1831)

183

BUTTERFLYFISHES • CHAETODONTIDAE

Beaked coralfish

Length: To 20 cm.
Distribution: Widespread tropical west Pacific, ranging to Andaman Sea.
Depth: 1-15 m.
General: Singly or in pairs on coastal or inner reefs and large open estuaries.
Feeds primarily on small invertebrates picked from narrow crevices.
A territorial species found abundant in some areas but rare or absent in central Indonesia.

Chelmon rostratus (Linaeus, 1758)

Very-long-nose butterflyfish

L: To 22 cm. Di: Indo-Pacific, throughout our area.
De: 5-60 m. Ge: Adults are nearly always in pairs and juveniles solitair.
Outer reefs along the deeper part of drop-offs. Some variations with darker shading and in some areas entirely dark brown.

Forcipiger longirostris (Broussonet, 1782)

Long-nose butterflyfish

Length: To 22 cm.
Distribution: Indo-Pacific, all of our area.
Depth: 3-30 m.
General: Usually in pairs. Exposed reefs on crests and upper sections of drop-offs.

Forcipiger flavissimus Jordan & McGregor, 1898

BUTTERFLYFISHES • CHAETODONTIDAE

Schooling Pyramid butterflyfish

Length: To 15 cm.
Distribution: West Pacific to Indian Ocean, in all of our area.
Depth: 5-60 m.
General: Along current swept drop-offs in coastal as well as outer reef zones, where they feed along the edge on plankton drifting past.
As its popular name says, it is mainly schooling.

Hemitaurichthys polylepis (Bleeker, 1857)

Schooling bannerfish
L: To 20 cm. Di: Indo-Pacific, in all of our area as Reef bannerfish H. *acuminatus* below (bottom). De: 1-100+ m.
Ge: A schooling species with thousands of individuals in current channels to feed on passing plankton. Juveniles are coastal and can be found on reef outcrops on mud or sand. They are also active cleaners sometimes approach divers where the activities take place.

Heniochus diphreutes Jordan, 1903

185

BUTTERFLYFISHES • CHAETODONTIDAE

Heniochus singularius Smith & Radcliffe, 1911

Singular bannerfish

Length: To 30 cm.
Distribution: Tropical west Pacific and east Indian Ocean, containing all of our area.
Depth: 2-50 m.
General: Adults nearly always in pairs on deep coastal and inner reef slopes, also seem to like shipwrecks. A rather shy species. Juveniles usually solitair in shallow lagoons but is also found in deeper water is small caves or ledges. Picks on the substrate for small invertebrate and algae.

Heniochus chrysostomus Cuvier, 1831

Pennant bannerfish

Length: To 16 cm.
Distribution: Widespread tropical west Pacific and east Indian Ocean, containing all of our area.
Depth: 1-40 m.
General: Adults usually in pairs along deep drop-offs but also found singly or in small groups in deep lagoons in clear coastal waters to outer reefs. Juveniles often very shallow in lagoons with mixed coral and algae reef. Their diet consists of invertebrates such as coral polyps, worms and crustaceans.

Horned bannerfish

L: To 20 cm. Di: West Pacific, west to Java. De: 2-30 m.
Ge: Occurs in pairs or in small groups on reef crests, usually in gutters or caves and prefers to remain in the shelter of the reef. Indian sibling, **Phantom bannerfish** *H. pleurotaenia* ranging east to Java, Indonesia (below).

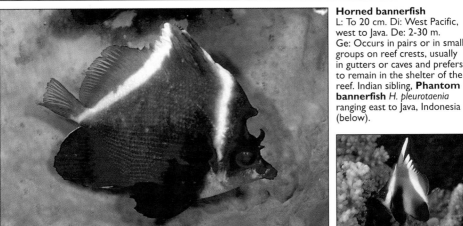

Heniochus varius (Cuvier, 1829)

ANGELFISHES • POMACANTHIDAE

This family comprises some of the most beautiful reef fishes, comparable to the butterflyfishes, however they are generally larger and more majestic. They are sought after by aquarist and highly prized. Over-exploiting species in some areas has led to banning imports of the entire family in many countries. They are readily distinguished from the closely related butterflyfishes by the large spine on the lower corner of the gill plate. The 7 genera with about 80 species comprises several distict groups: the larger species, of which *Pomacanthus* reaching 35 cm; the small species known as pygmy angels, *Centropyge*, some only reaching about 10 cm; and the planktivorous lyre-tailed *Genicanthus* its distinct shape. The larger species have the most dramatic different coloured juveniles imaginable, not even slightly relating to the adult, and *Genicanthus* has very different colour patterns between sexes. Juveniles of most species live secretive in reefs and are very territorial, particularly to their own kind but also to closely related species.

Except for *Genicanthus* they are benthic feeders, scraping and biting algaes of coral-rock, or sponges and corals, including the associated invertebrates, an activity which seems to take up most of the day. These fishes can produce low frequency drum-like or thumping noises, and when produced by the large species, so loud that it can startle a diver.

Angelfishes produce small round pelagic eggs. Hatchlings are less than 2 mm long and some attain 25 mm before settling, however most species settle on the substrate when about 10 mm.

A pair of **Pomacanthus navarchus**

ANGELFISHES • POMACANTHIDAE

Emperor angelfish

Length: To 38 cm.
Distribution: Indo-Pacific.
Depth: 6-60 m.
General: Adults usually in pairs, in deep lagoons with bommies and reef walls in caves, coastal as well as outer reefs.
Small juvenile (below) usually deep.

Pomacanthus imperator (Bloch, 1787)

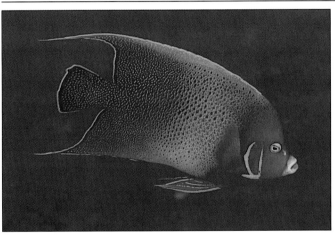

Half-circled angelfish

L: To 35 cm. Di: Indo-Pacific. De: 1-40 m. Ge: Coastal lagoons to outer reefs along drop-offs, but generally not common anywhere.
Juveniles (below) settle very shallow on rocky reef, secretive in small caves or ledges, feeding on short green algaes.

Pomacanthus semicirculatus (Cuvier, 1831)

Blue-ringed angelfish

L: To 45 cm. Di: Indonesia to southern Japan and to Africa. De: 1-60 m. Ge: A large and very impressive angel in our area, commonly found under jetties, in harbours and in shipwrecks. Adults usually in pairs in depths of 20 m or more. Juvenile below.

Pomacanthus annularis (Bloch, 1787)

ANGELFISHES • POMACANTHIDAE

Six-banded angelfish

L: To 46 cm. Di: Indo-Pacific, throughout our area.
De: 1-60 m. Ge: A large coastal species, adults usually in pairs. The juvenile (below) is difficult to distinguish from the next two species and best identified by the mainly white lines on the body and plain pectoral fin.

Pomacanthus sexstriatus (Cuvier, 1831)

Blue-face angelfish

L: To 36 cm.
Di: West Pacific and east Indian Ocean, throughout our area.
De: 5-45 m. Ge: With good coral growth but with many channels and caves. Adults mostly solitair and juveniles secretive on shallow reefs. Juveniles below.

Pomacanthus xanthometopon (Bleeker, 1853)

Majestic angelfish

L: To 25 cm. Di: Widespread tropical west Pacific.
De: 6-40 m. Ge: Outer reef drop-offs and deep lagoons in rich coral growth. Juveniles (see below) in outer reef lagoons and are best identified by the blue pectoral.

Pomacanthus navarchus (Cuvier, 1831)

ANGELFISHES • POMACANTHIDAE

Blue-striped angelfish

Length: To 22 cm.
Distribution: Tropical Japan, ranging south to Vietnam and Malaysia.
Depth: 10-60 m.
General: Mainly a common Japanese species which ranges far into the China Sea. Coastal rocky reefs along walls near ledges or caves. Juveniles secretive in ledges. Feeds mainly on sponges and tunicates.

Chaetodontoplus septentrionalis (Schlegel, 1844)

Vermiculate angelfish
L: To 18 cm. Di: Tropical west Pacific from Indonesia to southern Japan. De: 6-30 m. Ge: Inhabits rich coral growth areas of coastal reefs and lagoons. Adults nearly always in pairs, swimming close together. Eastward of Flores replaced by an unknown species with grey tail, photo below.

Chaetodontoplus mesoleucus (Bloch, 1787)

Phantom angelfish

Length: To 20 cm.
Distribution: Indonesia to Philippines.
Depth: 5-20 m.
General: Like all angelfishes this species uses cleaner stations by wrasses or shrimps to get rid of parasites.
Photo shows a cleaning *Lysmata amboinensis*.

Chaetodontoplus melanosoma (Bleeker, 1853)

ANGELFISHES • POMACANTHIDAE

Three-spot angelfish

Length: To 26 cm, usually to 20 cm.
Distribution: Indo-Pacific, throughout our area.
Depth: 3-60 m.
General: Coastal reef slopes on rich rubble ridges with soft corals, sponges and whips. Adults form small aggregations where common and juveniles secretive in rubble piles, especially with deep isolated outcrops with sponge growth in depths of about 30 m or more.

Apolemichthys trimaculatus (Lacépède, 1831)

Herald's dwarf-angelfish

Length: To 10 cm.
Distribution: West Pacific from southern Japan to Taiwan and eastern Australia.
Depth: 6-60 m.
General: A rare species in eastern Indonesia, only in very deep water on the foot of drop-offs along outer reefs. Usually in abundance where it occurs, busily moving through rubble reef. Only known from shallow depths in southern Japan. Sexual dimorphic with males developing pointed fins and a dark patch behind eye.

Centropyge heraldi Woods & Schultz, 1953

Lemon peel

L: To 14 cm, usually 11 cm.
Di: Two populations: Eastern Indian Ocean + Micronesia
De: 1-10 m Ge: A shallow living species with interesting distribution: The Indian Ocean population has no blue ring around the eyes, see large picture. Pacific form below.

Centropyge flavissimus (Cuvier, 1831)

191

ANGELFISHES • POMACANTHIDAE

Blue & gold angelfish

Length: To 16 cm, usually to 12 cm.
Distribution: West Pacific. Not known from Andaman Sea.
Depth: 3-20 m.
General: A commonly observed species in rich coral and rocky coastal reefs and lagoons. Adults often in small aggregations moving over a wide area of the reef whilst feeding, picking for a few seconds on algaes than quickly look for another place. Small juveniles hide and are rarely seen out in the open.

Centropyge bicolor (Bloch, 1787)

Damsel dwarf-angelfish

Length: To 8 cm.
Distribution: Eastern Indian Ocean to Philippines.
Depth: 10-60 m.
General: Occurs on rubble bottoms with spare coral growth. Common but very shy. Looks very similar to damselfishes because of its colouration and usually mistaken as such.

Centropyge flavicauda Fraser-Brunner, 1933

Moonbeam dwarf-angelfish

L: To 10 cm. Di: Sri Lanka to the Andaman Sea. De: 3-20 m. Ge: Similar to *C. multispinis* but can easily be distinguished by its yellow pectorals. Coral rubble slopes and mainly along bases of small drop-offs or along rich reef margins, usually solitair.

Centropyge flavipectoralis Randall & Klausewitz, 1977

ANGELFISHES • POMACANTHIDAE

Coral beauty

Length: To 10 cm.
Distribution: Indo-Pacific, throughout our area.
Depth: 3-60 m.
General: A common deep water species but also found in clear water lagoons in rich and dense coral growth. Highly variable from totally blue to red and very pale with thin vertical lines, usually in relation with depth and habitat. Usually blue with reddish sides. May pair or live in loose aggregations.

Centropyge bispinosus (Günther, 1860)

Rusty angelfish

Length: To 10 cm.
Distribution: Philippines to southern Japan.
Depth: 6-30 m.
General: Usually in depths of about 10 m, single or in small loose groups, feeding on filamentous algae from rocks or dead coral pieces. Variable in colour from a pale creamy colour to bright red.

Centropyge ferrugatus Randall & Burgess, 1972

Keyhole angelfish

L: To 15 cm. Di: Indo-Pacific, throughout our area.
De: 3-35 m. Ge: Coastal reefs, various habitats from rocky boulders to jetties. Small juveniles black except for white side spot which varies greatly in size. Large individuals obtain a blue sheen.

Centropyge tibicen (Cuvier, 1831)

ANGELFISHES • POMACANTHIDAE

Eibl's angelfish

L: To 10 cm. Di: Eastern Indian Ocean, ranging east in Indonesia to Flores. De: 3-25 m. Ge: Sibling of *C. vroliki* and hybrids are not uncommon. In rich coral growth. Juvenile (below) serve as model for the Indian Mimic Surgeon *Acanthurus tristis* (see photo).

Centropyge eibli Klausewitz, 1963

Pearly-scale dwarf-angelfish

Length: To 12 cm.
Distribution: Indo-Pacific, throughout our area.
Depth: 3-25 m.
General: Coastal reefs, commonly amongs rocky boulders with algae and sponge growth on which they feed. Less common in rich coral areas where they prefer the dead corals which are overgrown and this make them less conspicious. Juveniles secretive on protected shallow rocky reefs.

Centropyge vroliki (Bleeker, 1853)

Flame dwarf-angelfish

Length: To 10 cm.
Distribution: Northern Indonesia to the Philippines eastward.
Depth: 3-20 m.
General: Most attractive of all dwarf-angelfishes. Photo from Northern Sulawesi, Indonesia. Rare in our range, more common in oceanic locations.

Centropyge loriculus (Günther, 1874)

ANGELFISHES • POMACANTHIDAE

Regal angelfish

Length: To 25 cm.
Distribution: Indo-Pacific, throughout our area.
Depth: 3-80 m.
General: A common species in rich coral areas from shallow coastal reefs to outer reef drop-offs.
Small juveniles in caves along drop-offs. Some geographical variation: Indian Ocean form is yellow or orange on chest instead of grey of the west Pacific form.
Pictures right: Pacific with Indian below.
Below: juvenile and aberrant form from Indonesian reefs.

Pygoplites diacanthus (Boddaert, 1772)

Many-banded angelfish

Length: To 12 cm.
Distribution: Widespread west Pacific.
Depth: 3-70 m.
General: Secretive in caves and ledges along deep drop-offs. Sometimes in shallow reef cuttings. Often seen in small groups in large caves, upside-down on the ceilings. Small juveniles have a large ocellus posteriorly on dorsal fin.
This species is distinct from the dwarf-angelfish and is provisionally placed into this genus.

<Centropyge> multifasciatus (Smith & Radcliffe, 1911)

ANGELFISHES • POMACANTHIDAE

Blue-backed angelfish

L: To 12 cm. Di: Philippines north to Japan. De: 6-40 m. Ge: Prefers protected areas with clear water. Swims almost always upside-down in caves. Closely related to Many-banded angelfish, probably same genus. Hybrid from the Philippines below.

<Holacanthus> venustus Yasuda & Tominaga, 1969

Black-spot angelfish

L: To 18 cm. Di: West Pacific, Indonesia and east in our area. De: 20-50 m. Ge: Only seen on outer reef drop-offs with rich growth. Often in small aggregations dominated by the male. Sexual dimorphic as photographs show (female below)

Genicanthus melanospilos (Bleeker, 1857)

Lamarck's angelfish

L: To 22 cm. Di: Northern Australia, through Indonesia to southern Japan. De: 3-50 m. Ge: Unlike most angelfish, this species swims often midwater above reefs, especially the male in pursued of females which are loosely spread over the reef area. Juvenile below.

Genicanthus lamarck (Lacépède, 1802)

DAMSELFISHES • POMACENTRIDAE

Damselfishes or Demoiselles are a very large family, particularly in tropical coral reef habitats, and some species occur in such great numbers that they are probably the most numerous fishes in those areas. An estimated 300 species are globally distributed in tropical and sub-tropical waters. The species included here belong in 3 sub-families: AMPHIPRIONINAE, comprising the anemonefishes; CHROMINAE, the genera *Chromis, Acanthochromis & Dascyllus;* and POMACENTRINAE with the remaining genera. Most genera are distict in shape and other characters, but within a genus many species are similar and have geographical variations. Juveniles are often very different from the adult, whilst differences between sexes are small and often colour changes are only during the spawning period.

Habitats are nearly always reefs which feature plenty of cover such as small crevices and close to the food source. Basically, algae feeders occupy the shallow reef flats and planktivores the rocky outcrops. Only a few species move about over open substrates. Spawning varies from pair to community spawning in large groups. The demersal eggs vary in shape from ovoid to eliptical and in size from about 1-3.5 mm long. Hatching is often a few days and only one species has non-pelagic young: *Acanthochromis polyacanthus*. Post larvae settle at various sizes from about 10-20 mm depending on the species. Some juveniles have an extended pelagic by seeking shelter under floating weeds.

Damselfishes are popular with aquarists, in particular the anemone or clown fishes which are not only very colourful but have a close relationship with large anemones. IN the wild these fishes cannot survive without the protection of the stinging anemone. The anemone fishes live amongst the tentacles without getting stung, the anemone reacting to the fish contact like to other tentacles. These fishes are ideal for the aquarium as their territory in the wild is small, not expanding much beyond the anemone, and many species have been bred in captivity. In some areas the anemone's survival depends on the anemone fishes. If collectors remove all from the anemone, butterflyfishes may attack and eat the anemone. Some fish always should be left, preferably the larger breeding individuals.

Previous page:
A large colony of black anemonefishes *Amphirion melanopus,* typical for this species, photographed in the Banda Sea, Indonesia.

Premnas biaculeatus (Bloch, 1790)

Spine-cheek anemonefish

L: To 14 cm. Di: Widespread west Pacific and east Indian Ocean. De: 3-15 m. Ge: Protected coastal reefs, only found in the anemone *Entacmaea quadricolor.* Male small, female usually dark but orange in some areas. Obvious spine crossing white head band.

ANEMONEFISHES • POMACENTRIDAE

Tomato anemonefish
L: To 12 cm. Di: Andaman Sea, west Malaysia, east to Java. De: 3-15 m. Ge: Associate with two anemones: *E. quadricolor* and *H.crispa,* as shown. Juveniles have a white head band. Usually in pairs, but an individual was observed moving over a wide area between unoccupied anemones.

Amphiprion ephippium (Bloch, 1790)

Bridled anemonefish

L: To 14 cm. Di: West Indonesia, Malaysia, Thailand, Vietnam, Philippines.
De: 2-18 m. Ge: Adults with a single white bar, juveniles with two or three. Similar to *A. melanopus,* but has red ventral and anal fins. Associates only with *E. quadricolor.*

Amphiprion frenatus Brevoort, 1856

Black anemonefish

L: To 11 cm. Di: Eastern Indonesia, from Bali to Micronesia. De: 3-15 m. Ge: Associates mainly with *E. quadricolor* but also found in *H. crispa* & *H. magnifica.* In large numbers where anemones congregate. See also page 197.

Amphiprion melanopus Bleeker, 1852

ANEMONEFISHES • POMACENTRIDAE

Amphiprion chrysopterus Cuvier, 1830

Orange-fin anemonefish

Length: To 14 cm.
Distribution: Philippines, north-eastern Indonesia, ranging to Micronesia and central Pacific islands.
Depth: 3-30 m.
General: Associates with a large variety of anemones: *Entacmaea*, *Heteractis* & *Stichodactyla* genera, and the colour of the fish varies slightly in relation to different anemone genera. The pectoral fin is always orange or yellow but the ventral fins vary from orange to black. Clear water reefs, often in surge zones and found mainly in pairs.

Amphiprion sebae Bleeker, 1853

Seba anemonefish

Length: To 14 cm.
Distribution: Andaman Sea to Java, Indonesia in our area.
Depth: 1-10 m.
General: Highly variable, some individuals are entirely dark brown to blackish on the body (except for white bars), lacking yellow colour on the snout, breast and belly. It is known to have only one anemone host: *Stichodactyla haddoni*.

Pink anemonefish

L: To 10 cm. Di: Thailand, Malaysia, Indonesia, Philippines, Vietnam. De: 3-15 m.
Ge: Widespread in our area, even found at Cocos Keeling and Christmas Island in the Eastern Indian Ocean. Living together with various hosts. Distinct species with head band.

Amphiprion perideraion Bleeker, 1855

ANEMONEFISHES • POMACENTRIDAE

Western skunk anemonefish

Length: To 11 cm.
Distribution: In our area Andaman Sea (right) to Bali/Indonesia (below).
Depth: 3-15 m.
General: Shows a relatively narrow white stripe from top of the head to beginning of the dorsal fin, and continuing along the base of the fin in its entire length, which is broad in the next species below.
It is not known to occur together with this species but in Bali it co-occurs with another similar species, *A. perideraion*, and hybrids are common there.

Amphiprion akallopisos Bleeker, 1853

Eastern skunk anemonefish

Length: To 12 cm.
Distribution: In our range from east Indonesia to Philippines, the Solomon Islands and to West Australia.
Depth: 6-20 m.
General: Associates with *Heteractis crispa* & *Stichodactyla mertensii*.
Adults in pairs, often with many juveniles sharing anemone or in neighbouring ones. Coastal protected reefs prone to moderate currents.
In contrast to the western skunk anemonefish *Amphiprion akallopisos* tends to be orange, and its mid-dorsal stripe reaches the upper lip.

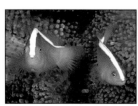

Amphiprion sandaracinos Allen, 1972

201

ANEMONEFISHES • POMACENTRIDAE

Clark's anemonefish

Length: To 14 cm.
Distribution: Widespread in our area: From Thailand to Philippines.
Depth: 1-20 m.
General: Usually black with variable amount of orange on head, ventral parts and fins. Three white bars on head, body and base of caudal fin. Photo location was in the Andaman Sea attracted by the unusual green color of the host *Stichodactyla haddoni*. Much less host-specific than other anemonefishes, known to occur in ten different anemone species!

Amphiprion clarkii (Bennett, 1830)

Panda clownfish

Length: To 13 cm.
Distribution: West Pacific, ranging to Singapore and south Java.
Depth: 3-35 m.
General: Associated with *Heteractis crispa* & *Stichodactyla haddoni*, the latter often well away from reefs on sandy or muddy substrates including deep channels between reefs. Highly variable within local groups and geographically. Also known as Saddleback anemonefish which could be applied to many species, adapted here the common name used in New Guinea.
The pair in the photograph are laying eggs and adopted a soft-drink-can for a nesting site.

Amphiprion polymnus (Linnaeus, 1758)

ANEMONEFISHES • POMACENTRIDAE

Western clown-anemonefish

Length: To 8 cm.
Distribution: West Pacific, except New Guinea and east Indian Ocean.
Depth: 1-15 m.
General: A common species in our area, associating with *Heteractis magnifica, Stichodactyla gigantea* & *S. mertensii*. Coastal protected reefs and usually in small groups.
Replaced by the next species further east, though they overlap in range between Sulawesi /Indonesia and New Guinea.

Right: Unusual variation in Andaman Sea.
Below: Normal colouration in most of our area.

Amphiprion ocellaris Cuvier, 1830

Eastern clown-anemonefish

Length: To 8 cm.
Distribution: West Pacific from north-eastern Australia and Vanuatu, north New Guinea coast as far as Sulawesi /Indonesia. Depth: 1-10 m.
General: As its western sibling species, associating with *Heteractis magnifica, Stichodactyla gigantea* & *S. mertensii*. Coastal protected reefs and usually shallow in small groups. Highly variable, especially when in *Stichodactyla* anemones, with black colouration.
The pictures included here are from New Guinea but this species was also photographed in Tomini Bay, Sulawesi in an atol lagoon.

Amphiprion percula (Lacépède, 1802)

HUMBUGS • POMACENTRIDAE

Black-tail humbug

L: To 80 mm. Di: West Pacific from northern Australia to Philippines. De: 1-10 m. Ge: Protected coastal reefs and lagoons. Small groups with *Acropora* corals as in the photograph. Similar common Humbug, *D. aruanus,* without black tail (below).

Dascyllus melanurus Bleeker, 1854

Indian humbug

L: To 60 mm. Di: Indian Ocean to Java, Indonesia. De: 3-30 m. Ge: The similar Humbug *D. reticulatus* below, replacing it further east throught the tropical west Pacific.
Both in small to large groups with *Acropora* heads on rock or near dead coral flats.

Dascyllus carneus Fischer, 1885

Three-spot dascyllus

L: To 80 mm. Di: Indo-Pacific, throughout our area.
De: 1-50 m. Ge: Juveniles often in great numbers with anemones, sometimes sharing with anemonefishes. Shallow reef tops to remote bommies in open substrate.

Dascyllus trimaculatus (Rüppell, 1828)

PULLERS • POMACENTRIDAE

Blue-green puller

L: To 85 mm. Di: Tropical Indo-Pacific. De: 1-20 m. Ge: Inhabits lagoons and coastal reefs in small to moderate sized schools, feeding closely above substrate on zooplankton. Similar **Green puller** *C. viridis* below.

Chromis atripectoralis Welander & Schultz, 1951

Golden sergeant

Length: To 12 cm.
Distribution: West Pacific and eastern Indian Ocean, throughout our area.
Depth: 6-30 m.
General: Coastal to outer reefs, in seawhip areas on which eggs are laid and defended by the male.
Easily recognise species by the all yellow colouration and deep body shape. Little variation with growth or between the sexes.

Amblyglyphidodon aureus (Cuvier, 1830)

Black & Yellow damsel

L: To 12 cm. Di: Widespread tropical west Pacific and east Indian Ocean to Andaman Sea. De: 3-30 m. Ge: Coastal to outer reefs in crest channels and rich coral slopes, areas prone to moderate currents. Population in Java and further east almost black, with more pointed fins. Juvenile below.

Neoglyphidodon nigroris (Cuvier, 1830)

DAMSELS • POMACENTRIDAE

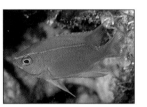

Royal damsel

Length: To 15 cm.
Distribution: Indo-Pacific, all of our area.
Depth: 1-15 m.
General: Juvenile (shown) solitair on coastal reef crests. Adult totally black, often in small loose groups.
Below, the juvenile Orange damsel, *N. crossi* looses the orange at early age, adult also drab, which is common in the Manado area of Indonesia. It ranges south to Flores and north to Philippines.

Neoglyphidodon melas (Cuvier, 1830)

Spiny-tail puller

Length: To 14 cm.
Distribution: Indonesia, Phillipines to Vanuatus and Great Barrier Reef.
Depth: 2-40 m.
General: Clear shallow coastal lagoons to deep outer reef drop-offs. Extremely variable between different areas from totally dusky to pale green or posteriorly white. Unique among damsels in taking care of young by guarding them closely (see photograph).

Acanthochromis polyacanthus (Bleeker, 1855)

SERGEANTS • POMACENTRIDAE

Sergeant Major

L: To 15 cm. Di: Widespread tropical Indo-Pacific. De: 1-15 m. Ge: Community spawners with numerous pairs producing eggs in circular patches over cleared rock or other flat surfaces. Below the Scissor-tail sergeant *A. sexfasciatus* another common species.

Abudefduf vaigiensis (Quoy & Gaimard, 1825)

Black-tail sergeant

Length: To 15 cm.
Distribution: Mainly eastern Indonesia from Flores to Philippines and to Solomon Islands.
Depth: 1-10 m.
General: Coastal reefs near mangroves and brackish zones with rocky reefs, likes man-made break waters and harbours.
Singly or in small loose groups, usually in a few metres depth. Juveniles in large rock pools.

Abudefduf lorenzi Hensley & Allen, 1977

Yellow-tail sergeant

Length: To 17 cm.
Distribution: Indo-Pacific, throughout our area.
Depth: 1-30 m.
General: Various shallow coastal habitats from high energy to quiet areas near freshwater run-offs, but also in high current channels such as the Komodo area where it schools in deep water.
Changes colour to a pale grey when swimming away from the substrate.

Abudefduf notatus (Day, 1869)

DAMSELS • POMACENTRIDAE

Yellowtail blue-damsel
L: To 50 cm. Di: Philippines, Indonesia, New Guinea, Solomon Islands. De: 1-15 m. Ge: Quiet coastal bays in dense and tall hard corals, often numerous on reefs. Two species mascarate under the name, as shown in photographs. Both occur in the Philippines, the type locality, and are widely distributed.

Chrysiptera parasema (Fowler, 1918)

Bleeker's damsel

Length: To 65 cm.
Distribution: Indonesia and Philippines.
Depth: 6-30 m.
General: Steep current prone coastal rubble slopes with sparse coral and algae growth, singly or few individuals together with small rock or coral outcrops. Only known from a few localities but may prefer deep habitats in other areas.

Chrysiptera bleekeri (Fowler & Bean, 1928)

Greenish damsel
L: To 80 mm. Di: From Bali to Thailand and to Philippines. De: 1-15 m. Ge: Coastal reefs dense coral cover, schooling in great numbers above when feeding on zooplankton. In Java most were near large plate corals in shallow depths.
Below the similar *N. cyanosomos* another common species.

Neopomacentrus anabatoides (Bleeker, 1847)

DAMSELS • POMACENTRIDAE

Philippine damsel

Length: To 10 cm.
Distribution: Widespread west Pacific and eastern Indian Ocean.
Depth: 1-15 m.
General: Highly variable species with different forms in different areas. The specimen in the photograph is typical for Indonesia. Coastal to outer reef crests on walls of shallow drops in cuttings and channels, usually solitair.

Pomacentrus philippinus Evermann & Seale, 1907

Princess damsel

Length: To 85 mm.
Distribution: Widespread west Pacific and eastern Indian Ocean.
Depth: 3-25 m.
General: Various habitats from coastal reef crests to lagoons and outer reef drop-offs. Highly variable in colour with habitat differences and geographically. Usually solitair, close to substrate.

Pomacentrus vaiuli Jordan & Seale, 1906

Yellow-belly damsel

L: To 10 cm. Di: Indonesia.
De: 2-15 m. Ge: Only recently separated as a valid species from more common *P. coelestis* which is blue on the belly as well. Below *P. alleni* from Andaman Sea ranging to south Java where it is very common.

Pomacentrus auriventris Allen, 1991

DAMSELS • POMACENTRIDAE

Yellow damsel

Length: To 80 mm.
Distribution: Widespread west Pacific and east Indian Ocean.
Depth: 1-15 m.
General: Sheltered coastal reefs and lagoons to outer reef habitats. Rich coral areas, corals used for shelter. Usually in small and loose aggregations. Variable from orange to bright pale yellow in our area.

Pomacentrus moluccensis Bleeker, 1853

Jewel damsel

Length: To 10 cm.
Distribution: Widespread tropical Indo-Pacific.
Depth: 3-15 m.
General: Coastal reefs, lagoons and outer reefs. In mixed coral and algae habitat.
Very territorial and aggressive towards many other fish species. Small juveniles with many irridescent blue centred ocelli over body which gradually fade with growth and are lost in adults.

Plectroglyphidodon lacrymatus (Quoy & Gaimard, 1825)

Dick's damsel

Length: To 10 cm.
Distribution: Widespread tropical Indo-Pacific.
Depth: 1-15 m.
General: Clear water coastal reefs, lagoons, to outer reefs in large coral heads.
Often in small loose aggregation comprising various sizes.
A distinct species, showing little variation with growth or between different areas.

Right: Normally seen with black coral in deep water, the Longnose hawkfish (page 214) occurs in our area also in shallow water.

Plectroglyphidodon dickii (Liénard, 1839)

HAWKFISHES • CIRRHITIDAE

The hawkfishes are a tropical family, 9 genera and 35 species are known, mostly distributed in the Indo-Pacific and only 3 species occur in the Atlantic seas. These small fishes hug the bottom, perched on the thickened lower pectoral fin rays, but unlike most bottom dwellers they are very active, often restlessly moving positions. Only one species regularly swims above the substrate to feed on zoo-plankton. Most species live in the shallows, on reef crests and strong surge zones, some can be found deeper than 30 m. Most species are habitat specific, found with certain sponges or corals and usually occur in loose aggregations. They are carnivores, feeding on small fishes and invertebrates, generally small and less than 10 cm long but a few attain almost 30 cm.

Hawkfishes are pelagic spawners, post larvae are fairly large and usually in excess of 20 mm and up to about 40 mm long. Filaments at the tip of the dorsal spines, often tufted, are diagnostic for the family.

Cirrhitichthys aprinus (Cuvier, 1829)

Blotched hawkfish

Length: To 12 cm.
Distribution: Widespread tropical west Pacific, including all of our area.
Depth: 5-40 m.
General: Blotched pattern variable, in some populations forming bars.
Mainly on coastal reefs and rocky estuaries, usually with sponges. In pairs or small aggregations, common in some areas.

Cirrhitichthys aureus (Temminck and Schlegel, 1843)

Golden hawkfish

Length: To 15 cm.
Distribution: Tropical northwest Pacific, ranging from southern Indonesia, Philippines to Japan.
Depth: 20-60 m.
General: Colour yellow to brown, sometimes with dark blotches similar to *C. aprinus*. On sponges along outer reef drop-offs, usually in depths of 20 m or more and not common anywhere known, only observed singly.

HAWKFISHES • CIRRHITIDAE

Coral hawkfish

Length: To 10 cm.
Distribution: Widespread tropical west Pacific, containing all of our area.
Depth: 5-40 m.
General: Clear coastal reef crests and rubble slopes to deep outer reef habitats. Juveniles solitair, adults usually in pairs. Colour slightly varies between habitat with pale specimens from white sand reefs to deeply colourations on darker reefs.

Cirrhitichthys falco Randall, 1963

Spotted hawkfish
L: To 10 cm. Di: Widespread tropical Indo-Pacific, containing all of our area. De: 3-40 m. Ge: Shallow coastal rocky-boulder reefs to deep rubble slopes. Usually observed singly. Colouration varies from dark to light to suit various habitats, sometimes bright red in deep habitats.

Cirrhitichthys oxycephalus (Bleeker, 1855)

Lyre-tail hawkfish

Length: To 10 cm.
Distribution: Widespread tropical Indo-Pacific, including all of our area.
Depth: 15-50 m.
General: Coastal to outer reef slopes, often in large aggregations with remote outcrops of rock or coral, in current prone areas. Swims well above substrate, often mixed with basslets to feed on zooplankton. Colour varies from light to dark brown.

Cyprinocirrhites polyactis (Bleeker, 1875)

HAWKFISHES • CIRRHITIDAE

Ring-eyed hawkfish

Length: To 14 cm.
Distribution: Widespread tropical Indo-Pacific, throughout our area.
Depth: 1-10 m.
General: Mainly on shallow outer reef crests, but also clear coastal reefs, perched between outer coral branches. Often in small aggregations per coral head and numerous individuals in neighboring corals. Variable dark brown to grey and red, white band on side sometimes indistinct.

Paracirrhites arcatus (Cuvier, 1829)

Forster's hawkfish
L: To 20 cm. Di: Indo-Pacific, throughout our area.
De: 3-30 m. Ge: Clear coastal to outer reef slopes, usually perched among the outer-most coral branches to observe surroundings. Sometimes all dark with yellowish tail and adults with additional spotting on head.

Paracirrhites forsteri (Schneider, 1801)

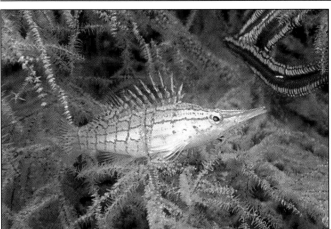

Longnose hawkfish

Length: To 10 cm.
Distribution: Indo-Pacific, including all of our area.
Depth: 5-70 m.
General: Primarily a deep water dweller but in some areas as shallow as 5 m, where black coral or gorgonians grow which they seem to associate with. See also page 211.

Oxycirrhites typus Bleeker, 1857

BANDFISHES • CEPOLIDAE

A little known family of small fishes which are widely distributed along continental margins of the western Pacific, Indian Ocean and eastern Atlantic. Comprises at least 4 genera and about 20 species. A study of the Pacific species is needed. Many similar, eel shaped, fishes with a tapering body and dorsal and anal fins along almost the entire length, and seemingly continuous with the pointed caudal fin. Rounded goby-like head with large mouth and large eyes. Most species with identical colours, usually red or pink. Primarily they live in muddy or fine-sand substrates in self-made burrows, in a depth range of 20-500 m, feeding on zoo-plankton. Comes out the burrows during currents, hoovering vertically and looking upwards for food. Returns to burrow by swimming backwards. Groups feeding form an interesting sight in an otherwise barren environment which may be shared by a few gobies or eels. The burrows are straight down and open on top. Somehow the fish manages the mud or sand from falling or cascading down, perhaps using slime or mucus which hardens. Each fish has ist own burrow and during the feeding sessions the males often display to the nearby females.

Although no spawning has been observed, eggs of some species are known to be pelagic.

Scarlet bandfish

Length: To 20 cm.
Distribution: Indonesia +?
Depth: 6-40 m.
General: A common species in Flores and also observed in Milne Bay, Papua New Guinea, no doubt widespread throughout our area. Occurs in small to large agregations on mud or sand falts and slopes, coming out to feed when the currents cary the zooplankton to feed on. Several similar species and several names available for this species.

Cepola sp

JAWFISHES • OPISTOGNATHIDAE

A family of small burrowing species, many of which are undescribed, comprising 3 genera and an estimated 70 species. Distributed in all tropical waters, but the east Pacific. Somewhat odd-looking fishes with large eyes, placed high and forward on head, a large mouth used for incubating eggs, and elongate bodies. The burrows are usually vertically in the sand, and are reinforced with small rocks or coral bits and looking down in it is somewhat like a wishing-well. These curious fishes are given a variety of common names, including Missing Links, Pugs, Harlequins, Grinners, Smilers, and Monkeyfish. The ventral fins are peculiar in having outer two rays unbranched. Usually they are found with the head just out of the borrow, watching out for food floating past, zooplankton, and retreat to burrow is tail first. Sometimes shrimps share their burrows.
Some species live in colonies with burrows evenly dotted over a suitable area, usually rubble areas surrounded by reef. Steeling of building material is not uncommon and so are the disputes over them. An interesting family for the aquarists, but much work is needed to put even one name to the species in our area.

JAWFISHES • OPISTOGNATHIDAE

Opistognathus sp

Gold-specs jawfish
L: To 11 cm. Di: Indonesia, Philippines. De: 3-20 m.
Ge: A common species from Bali to Flores. Sometimes has unidentified pontoniine shrimp as shown in left photograph, but relationship is not clear. The shrimp is known from Flores to Andaman Sea, shown with next species.

Opistognathus sp

Blue-spotted jawfish
L: To 12 cm. Di: Only known from Similan Island, Thailand. De: about 25 m.
Ge: On open rubble. A pair was found and some more specimens discovered nearby one together with a shrimp of the genus *Palaemonella*. Like most species in our area, it appears to be undescribed.

Opistognathus sp

Variegated jawfish
L: To 11 cm. Di: Indonesia, Bali to Flores. De: 10-50 m.
Ge: A variable species in colour from bright yellow to brown, seemingly representing a single one, perhaps is **Opistognathus solorensis** Bleeker. Another apparently undescribed species is Ring-eye jawfish below from the same areas in Indonesia.

MULLETS • MUGILIDAE

A large family with approximately 13 genera containing an estimated 70 species which are distributed in all but the coldest seas, estuaries and some species enter fresh water. Mullets have moderately elongate bodies and a broad depressed head. Two widely separated dorsal fins which are of similar size. Mouth small, terminal and jaws usually with or without loosely attached teeth in one group and firmly attached more rigid teeth in another, which are regarded as sub-families.

Most species occur in large schools along the coast and migrate to spawn, some into estuaries and others out to sea. They feed by taking materials of rocks or mud, either pre-filtered through the gills or swallowed. Sometimes large quantities of nutriment rich mud is swallowed for treatment with the specially adapted stomach which has a muscular portion similar to the gizzard of a bird.

Warty-lip mullet

Length: 40 cm.
Distribution: Tropical Indo-Pacific, including all of our area, ranging to subtropical waters.
Depth: 1-20 m.
General: Silvery with grey longitudinal lines along scale rows, slightly darker, sometimes greenish above. Usually roaming in large schools over shallow sandy slopes. May swim just below the surface until the leader decides it's feeding time and all follw down to the substrate to scope up the upper layer of the sand to filter feed.

Crenimugil crenilabrus (Forsskal, 1775)

Diamond-scale mullet

Length: 50 cm.
Distribution: Tropical Indo-Pacific, including all of our area.
Depth: 0.1-6 m.
General: A common estuarine species found along beaches on shores and in lagoons. Juveniles swim in schools against the surface and have pretty black and yellow fins. Adults feed on the shallow sand substrate below. A distinct species by its large scales and dark pectoral fins.

Liza vaigiensis (Quoy & Gaimard, 1824)

WRASSES • LABRIDAE

The wrasses make up one of the largest families of fishes on tropical reefs, worldwide it comprises more than 60 genera and an estimated 400 species. The various genera can be divided into several distinct groups but their exact relationship is not clear. Some genera too have species groups which may represent subgenera or another valid genus. Much research is needed to clearify the present situation. About 150 species occur in Indonesian waters, many of which recently described and some are yet to be named. This number will probably increase as exploration to new areas and investigations to the family continues.

Wrasses are highly diverse in shape and range greatly in size. Nearly all wrasses have various and distinct colour forms during growth, from post-larvae-stage to adult, and between sexes. Normally juveniles become female first and males derive from females. In almost all species the fully developed males become very colourful and can look totally different from the earlier stages. Most males dominate a number of females and are territorial towards other males. If a male disappears from a group, the next most dominant adult, usually largest and normally still female, takes over and changes sex. They are mostly benthic feeders, taking a great variety of invertebrates but some prefer small fishes or eggs. A few are specialized feeders, taking plankton or pick parasites from other fishes. All species are diurnal and retire by either burying in the sand or by wedging themselves into crevices. Wrasses are primarily shallow reef dwellers, but a few species have adapted to open sand or rubble and may occur deep.

All the Indo-Pacific wrasses are thought to produce pelagic eggs. Post-larvae settle at various sizes, depending on species, from about 6-15 mm. Sand burying wrasses settle on sand patches among reef or close to small outcrops of rock or seagrass in open sand areas. The non-burying species settle secretive among rocks and often in the protection of long-spined urchins.

Wrasses are easily kept in aquaria, however some species have special requirements. The small species which like to burry in the sand do not like coarse substrate often used, whilst the large species need space. It is interesting to grow small juveniles to adult stages and watch the changes taking place, sometimes changing to male. The latter usually takes only about a week after the first sight of colour change. The small planktivores such as *Cirrhilabrus* and *Paracheilinus* are ideal for the invertebrate aquarium.

Lyretail hogfish
L: To 22 cm. Di: Indo-Pacific, throughout our area.
De: 6-60 m. Ge: Coastal to outer reefs in rich coral growth along drop-offs. Usually depths in excess of 20 m but if suitable habitat much shallower, seen in 6 m with large gorgonian fans. Juveniles may clean other fishes from parasites.

Bodianus anthioides (Bennett, 1830)

WRASSES • LABRIDAE

Blackbelt hogfish

L: To 20 cm. Di: Widespread tropical west Pacific and eastern Indian Ocean. De: 5-40 m.
Ge: Mostly found in clear outer reef waters in rich coral growth, but in some areas they are just as common on coastal reefs. Juveniles dark brown with large yellow spots, secretive in caves. Juvenile of the similar *B. axillaris* has white spots, both shown below.

Bodianus mesothorax (Schneider, 1801)

Diana's hogfish

Length: To 24 cm.
Distribution: Indo-Pacific, two forms (species?) in our area.
Depth: 3-40 m.
General: Indian Ocean form, ranging to Java, in which the adults of the former looses the black blotches on the ventral and anal fins, becoming plain red (below). Pacific form ranges west to Java and it retains the spots in the fins. Outer reefs in rich coral growth, particularly soft corals, from shallow reef flats to deep drop-offs. Juveniles usually deep in about 20-30 m, in soft corals or swimming upside down on the ceiling of large overhangs, sometimes in small aggregations.

Bodianus diana (Lacépède, 1801)

219

WRASSES • LABRIDAE

Bodianus loxozonus (Snyder, 1908)

Eclipse hogfish

Length: To 40 cm.
Distribution: North eastern Indonesia to southern Japan, and eastern New Guinea to Great Barrier Reef.
Depth: 3-40 m.
General: Occasionally seen on outer reefs crest of oceanic localities, mainly north-eastern Indonesia. Juveniles similar to adult but lines formed by small spots and larger posterior black area on body. The similar *B. macrourus* replaces this species in the Indian Ocean and may occur in areas of south-western Indonesia.
It differs in colour by having a broad black band posteriorly.

Bodianus bilunulatus (Lacépède, 1801)

Saddleback hogfish

Length: To 55 cm, but usually to 35 cm.
Distribution: Indo-Pacific, throughout our area.
Depth: 10-100 m.
General: Slight differences between west Indian Ocean and west Pacific in face markings and size of black saddle on adults.
Adults solitair on deep coastal sand and rubble slopes with large coral heads. Juveniles on rubble ridges with sponge and soft coral growth.

Bodianus bimaculatus Allen, 1973

Yellow hogfish

Length: To 10 cm, usually to 75 mm.
Distribution: Indo-Pacific, throughout our area.
Depth: 20-60 m.
General: Very deep along drop-offs to at least 60 m, but occurs in Bali on rich sand slopes, comprising boulder rock covered with soft corals and other invertebrates, as shallow as 20 m.
Usually in small aggregations and in Bali juveniles often mixed with the bright yellow *Halichoeres chrysus* juveniles. The smallest *Bodianus*.

WRASSES • LABRIDAE

Cigar wrasse

Length: To 48 cm.
Distribution: Indo-Pacific, throughout our area.
Depth: 1-30 m.
General: Colour varies little between stages, but sometimes a bright yellow form. Seagrass beds and weed areas, but also on reefs or soft corals gardens where they are often yellow. Shallow depths but observed on reef slopes with soft corals to about 30 m.

Cheilio inermis (Forsskal, 1775)

Banded thicklip

Length: To 50 cm, usually to 30 cm.
Distribution: Indo-Pacific, throughout our area.
Depth: 1-40 m.
General: Post-larvae are very small, only about 5 mm total length and usually found amongst sea-urchin spines. Juveniles on coastal reefs and in lagoons on coarse rubble patches, avoiding exposure by swimming through passages in the substrate. Outer reefs, shallow to about 40 m in lagoons and on rubble slopes.

Hemigymnus fasciatus (Bloch, 1792)

False leopard wrasse
L: To 13 cm. Di: Indonesia and Andaman Sea. De: 3-25 m.
Ge: Replaced by similar sibling species *M. meleagris*, the leopard wrasse (below), north of Manado and eastern most parts of Indonesia. Coastal reefs, outer reef crests and lagoons, usually in small aggregations comprising several females.

Macropharyngodon ornatus Randall, 1978

221

WRASSES • LABRIDAE

Dusky wrasse

L: To 16 cm. Di: Indo-Pacific, throughout our area.
De: 3-30 m. Ge: Coastal reefs from shallow reef flats and slopes to about 30 m. Small juveniles in crevices, often with sea urchins.
Below the common Hoeven's wrasse H. melanurus.

Halichoeres marginatus Rüppell, 1835

Chain-lined wrasse

L: To 14 cm. Di: Indonesia to southern Japan and New Guinea.
De: 3-15 m. Ge: The long snout makes this species distinctive. Soft coral gardens on reef flats, along upper edge of drop-offs. H. richmondi is a synonym. The similar H. purpurescens below.

Halichoeres leucurus (Walbaum, 1792)

Saowisata wrasse

L: To 24 cm. Di: Indonesia, Floris, Bali and northern Sulawesi possibly widespread.
De: 3-20 m. Ge: Coastal and protected inner reefs in bays and lagoons along short drop-offs, in some place sharing habitat with the closely related H. timorensis (below).

Halichoeres binotopsis (Bleeker, 1849)

WRASSES • LABRIDAE

Indian white wrasse
L: To 12 cm. Di: Widespread tropical Indian-Ocean to Indonesia. De: 20-60 m. Ge: Only recently found in southern Java on a rocky sea mount at 25 m depth, a small groups of about six headed by the male in the photograph. Replaced by sibling species further east. Below the similar *H. melasmopomus*.

Halichoeres trispilus Randall & Smith, 1982

Yellow wrasse

L: To 12 cm. Di: West Pacific to Java, Indonesia. De: 3-30 m. Ge: Rubble slopes and drop-offs, usually in small groups with male in charge. Replaced by its sibling, *H. leucoxanthus* in eastern Indian Ocean and both occur in Java.

Halichoeres chrysus Randall, 1980

Blue-ribbon wrasse
L: To 14 cm. Di: West Pacific. De: 3-20 m. Ge: Mainly along outer reefs in small to large aggregations of mixed sex. Males more obvious than females with the brighter colour and by often swimming high above substrate.
The common **Silver-streaked wrasse** *S. strigiventer* below.

Stethojulis triliniata (Bloch & Schneider, 1801)

223

WRASSES • LABRIDAE

Speckled wrasse
L: To 21 cm. Di: Indo-Pacific. De: 3-30 m. Ge: Coastal slopes and outer reefs crests and drop-offs to moderate depths. Juvenile solitair, adults usually in loose pairs or small groups of females with male travelling a large reef section. Female below.

Anampses meleagrides Valenciennes, 1839

White-dashed wrasse
L: To 12 cm. Di: Indian Ocean. De: 10-40 m. Ge: Coastal slopes and drop-offs in small aggregations, dominated by a single male, female below. Its sibling species, *Anampses melanurus* further east Both species occur in Bali and the latter differs mainly in having round spots instead of dashes.

Anampses lineatus Randall, 1972

Narrow-banded wrasse
L: To 36 cm. Di: Indo-Pacific. De: 3-40 m. Ge: Rubble reef flats and slopes. Found in small aggregations and often mixed with other species of the same genus. Males often seen solitair, but normally there are a number of females. Juvenile below.

Hologymnosus doliatus (Lacépède, 1801)

WRASSES • LABRIDAE

Gaimard's wrasse
L: To 35 cm. Di: West Pacific and east Indian Ocean.
De: 3-50 m. Ge: Mainly with rubble reef sections, shallow outer reef flats and slopes. Juveniles solitair amongst boulders. Adults solitair or in small loose aggregations. Often seen turning over rocks or coral pieces to expose prey. Various colour changes during growth. Below: Male and intermediate phase with yellow tail.

Coris gaimard (Quoy & Gaimard, 1824)

Right: Pixie wrasse

Length: To 12 cm.
Below: Batu wrasse *C. batuensis* and *C. dorsomacula*. These three species are common throughout our area, all living near sand patches or along reef fringes bordering onto sand. The pixie wrasse usually deeper than 20 m, the others shallow.

Coris pictoides Randall & Kuiter, 1982

225

WRASSES • LABRIDAE

Bird-nose wrasse
L: To 22 cm. Di: Widespread tropical west Pacific.
De: 3-20 m. Ge: Lagoons and reef crests in rich coral areas. Adults usually in small and loose aggregations dominated by large colourful male.
Below: female. Sibling Indian Ocean *G. caeruleus*, in Java and further west.

Gomphosus varius Lacépède, 1801

Six-barred wrasse

Length: To 20 cm.
Distribution: Indo-Pacific, throughout our area.
Depth: 3-15 m.
General: Juveniles in algae-reef habitat and near seagrasses. Adults coastal as well as along outer reefs near drop-offs and lagoons, shallow and sometimes intertidal. Small juveniles secretive in algae habitat, forming small aggregation with time and adults in moderate sized loose aggregations, often with several males.

Thalassoma hardwicke (Bennett, 1828)

Surge wrasse

Length: To 36 cm, usually to 24 cm.
Distribution: Indo-Pacific, throughout our area.
Depth: 0.5-10 m.
General: Coastal reefs and exposed reefs in shallow surge zones. Males patrol large sections of reefs and usually a few females are present, variously distributed over a large area claimed by the male.
The largest species, attains 36 cm in some areas (in cooler part of range).

Thalassoma purpureum (Forsskal, 1775)

WRASSES • LABRIDAE

Red-ribbon wrasse
L: To 15 cm. Di: Indo-Pacific.
De: 3-15 m. Ge: Coastal to outer reef crests, clear water and usually shallow to about 10 m. Juveniles solitair, females spread out on a section of reef, claimed by the male which dominates the others, including juveniles. Below: Male turning yellow during display.

Thalassoma quinquevittatum (Lay & Bennett, 1839)

Yellow moon-wrasse

L: To 25 cm. Di: West Pacific and east Indian Ocean.
De: 3-30 m. Ge: Most common towards the cooler part of the range. Outer reefs in lagoons and slopes to 30 m. Small juveniles solitair. Adults numerous in some areas. Below the common moon wrasse *T. lunare*.

Thalassoma lutescens (Lay & Bennett, 1839)

Jansen's wrasse

L: To 20 cm. Di: Indo-Pacific, two forms. De: 3-20 m.
Ge: Indian Ocean and Pacific have different colour forms, possibly valid species. Outer reef lagoons and slopes to about 20 m. Indonesian form shown, with female below.

Thalassoma janseni (Bleeker, 1856)

WRASSES • LABRIDAE

Razor wrasse

Length: To 20 cm.
Distribution: Indo-Pacific, throughout our area.
Depth: 3-20 m.
General: Males with a dark bar just behind head. Juveniles solitair and secretive in seagrasses or green weeds on sand. Adults in open patches but wary, quickly diving into sand, sometimes in pairs. Because of shy nature it is usually buried in the sand before a diver sees it.

Cymolutes torquatus Valenciennes, 1840

Rockmover wrasse
L: To 30 cm. Di: Indo-Pacific. De: 3-20 m. Ge: Juveniles on reef crests. Adults are experts in turning over large pieces of dead coral or rocks for a feed, working in pairs with one lifting and the other taking care of exposed prey, and taking turns. Large juvenile on left and small one below.

Novaculichthys taeniourus (Lacépède, 1801)

Blue razorfish
L: To 36 cm. Di: Indo-Pacific. De: 6-100 m. Ge: Mainly on sand slopes, or deep sand flats, juveniles relatively shallow but adults rarely in less than 30 m. Quickly dives in the sand when approached. Juveniles solitair and adults in loose groups, spread out over an area. Below: Male.

Xyrichtys pavo Valenciennes, 1839

WRASSES • LABRIDAE

White-blotch razorfish

Length: To 20 cm.
Distribution: West to central Pacific.
Depth: 1-25 m.
General: Adults typically as shown with variable intension of dark bands. Coast sand slopes with a preference to clean areas regardless of colour which they match by being very pale on white sand and dark banded on dark sand.

Xyrichtys aneitensis (Günther, 1862)

Red-spots razorfish
L: To 20 cm. Di: Indo-Pacific.
De: 1-20 m. Ge: Small juveniles, uniformly brown, white, bright yellow or green to almost black. Adults as shown in photographs. A common species throughout Indonesia on coastal mud and sand slopes. Juveniles solitair. Male with red spots, female shown below.

Xyrichtys pentadactylus (Linnaeus, 1758)

Tube-mouth wrasse

Length: To 20 cm.
Distribution: Indo-Pacific, throughout our area.
Depth: 3-20 m.
General: Ventral fins in male becomes greatly elongated, up to half body long. Sheltered coral gardens on reefs and rich lagoons in depths to about 20 m. Juveniles sometimes cleaning. Adults feed mostly on coral polyps. Pacific form shown, Indian Ocean form not as colourfull and grows larger than former which attains about 15 cm.

Labrichthys unilineatus (Guichenot, 1847)

WRASSES • LABRIDAE

Labrobsis xanthonota Randall, 1981

V-tail tubelip wrasse

Length: To 12 cm.
Distribution: Indo-Pacific, throughout our area.
Depth: 3-55 m.
General: Most distinctive feature in male is the unusual V colouration of the caudal fin. Mainly found on clear outer reef crests and the upper zone along drop-offs with good coral growth, usually shallow, but reported deep to 55 m in some areas.

Labroides dimidiatus (Valenciennes, 1839)

Cleaner wrasse

Length: To 10 cm.
Distribution: Indo-Pacific, throughout our area.
Depth: 2-30 m.
General: In some areas yellow over the back and usually this form is much deeper in about 20 m or more. Rocky and coral reefs from estuaries to off shore reefs.
Small juveniles singly, ranging to warm-temperate seas and interestingly activily cleans temperate species of fish. Adults usually in pairs.

Cirrhilabrus exquisitus Smith, 1957

Exquisite fairy-wrasse

Length: To 12 cm.
Distribution: Indo-Pacific, throughout our area.
Depth: 1-30 m.
General: Some geographical variation between different oceans (see Indian Ocean Guide, Debelius, 1993).
In small to large aggregations on protected outer reef crests and inner reef lagoons. Males usually busy displaying to each other.

WRASSES • LABRIDAE

Blue-head fairy-wrasse
L: To 13 cm. Di: Along mainland Asia from Andaman Sea to southern Japan. De: 1-35 m. Ge: Large males with blue-rimmed scales centrally on body. Coastal reefs in small to large aggregations, comprising young adults and numerous males. Large aggregations in Java. Below: male from Japan.

Cirrhilabrus cyanopleura (Bleeker, 1851)

Red-head fairy-wrasse

L: To 13 cm. Di: Indonesia, from Bali and Sulawesi to the Flores region. De: 3-20 m. Ge: Coastal reefs, shallow to depths of about 20 m, usually in small to large aggregations and usually colouful males greatly outnumbered by other phases. Female shown below.

Cirrhilabrus solorensis (Bleeker, 1853)

Yellow-fin fairy-wrasse

L: To 9 cm. Di: Indonesia, ranging to Philippines. De: 6-20 m. Ge: Prefers rich coral reefs in protected bays, often in large schools, in depths of about 6-20 m. Closely related to *C. lubbocki* (below) and sometimes the species mix.

Cirrhilabrus flavidorsalis Randall & Carpenter, 1980

WRASSES • LABRIDAE

Paracheilinus filamentosus Allen, 1974

Filamented flasher wrasse

Length: To 10 cm.
Distribution: Indonesia, Philippines to Solomon Islands.
Depth: 6-35 m.
General: Extremely variable between different localities and males change colour during display. Coastal and inner reef slopes in rich coral areas with rubble zones, usually in great numbers along slopes feeding high above the substrate on zooplankton. Large males boltly swim around, sometimes seemingly getting into a display frenzy with numerous males flashing there colours and fins.

Paracheilinus sp 1

Yellow-fin flasher wrasse

Length: To 8 cm.
Distribution: Indonesia, Bali to Flores along the north coasts.
Depth: 15-35 m.
General: Males with the first or first two soft rays extended prong-like in dorsal fin, which turn red during display. Large yellow anal fin also used for fisplay. Extremely variable between different localities and males change colour during display with bright red prongs. Deep coastal reef slopes in rich low reef zones, usually in small groups comprising juveniles, females and a single large male. Appears to be undescribed.

Paracheilinus sp 2

Pink flasher wrasse

Length: To 8 cm.
Distribution: Indonesia, Bali and Flores along north coasts.
Depth: 25-35 m.
General: Males usually with three or four prong-like extended rays in soft dorsal fin which turn red during display. Only seen in depths of 25 and 35 m. and appears to be rare. Small aggregations on deep rubble slopes comprising females with a single male. Probably undescribed.

WRASSES • LABRIDAE

Pin-striped wrasse

Length: To 8 cm.
Distribution: Indo-Pacific, throughout our region.
Depth: 10-40 m.
General: Coastal to outer reefs on rich boulder slopes and along drop-offs in rich sponge caves to moderate depths. Singly or in small loose aggregations. Several similar species, all secretive in shelter of narrow crevices of corals or rocks.

Pseudocheilinus evanidus Jordan & Evermann, 1903

Cockerel wrasse

Length: To 15 cm.
Distribution: West Pacific, probably all of our area.
Depth: 3-30 m.
General: Various habitats from shallow algae reefs to deep off shore on soft bottom with sponges and hydroid colonies. Usually hiding and only seen when moving to the next shelter nearby. Several can be seen in a relatively small area and probably are members of a small group.

Pteragogus enneacanthus (Bleeker, 1852)

White-banded possum-wrasse

Length: To 55 mm.
Distribution: West Pacific, Indonesia and Philippines but probably widespread.
Depth: 10-40 m.
General: Colour pattern changes little with size and much easier to identify from related species because of the angle of the white lines accross the body. Seems to prefer outer reef conditions and typically lives far in the back of caves and ledges along steep drop-offs.

Wetmorella albofasciata Schultz & Marshall, 1954

233

WRASSES • LABRIDAE

Slingjaw wrasse
L: To 35 cm. Di: Widespread our area. De: 3-40 m. Ge: Juv. often mistaken for *Wetmorella sp* because of its colouration. Coral reefs, coastal to outer inner reefs and usually solitair. Adults often shy, male left, female below. Presently a single species recognised but genus may comprise 3 species.

Epibulus insidiator (Pallas, 1770)

Slender splendour wrasse
L: To 15 cm. Di: West Pacific. De: 3-30 m. Ge: Typical maori scribbles on the face. From outer reef lagoons to deep slopes, with rich growth of corals or other invertebrates. Common in suitable habitat and groups widely dispersed over the reefs. The other common *C. unifasciatus* below.

Cheilinus celebicus Bleeker, 1853

Banded splendour wrasse

Length: To 35 cm.
Distribution: Indo-Pacific.
Depth: 6-40 m.
General: Juveniles distinctly banded, similar to ***Wetmorella*** when very small. Coastal to protected inner reefs, occasionally trawled in deep water. Juveniles secretive and solitair in dense corals. Adults in loose groups, often curious towards divers.

Cheilinus fasciatus (Bloch, 1791)

PARROTFISHES • SCARIDAE

A large tropical family with representatives in all oceans. Medium sized fishes which are characterized largely by their teeth, in 2 sub-families, one group the jaws with numerous conical teeth, the other fused teeth into plates, beak-like similar to parrots, hence their common name. In all 9 genera, and an estimated 80 species.

The majority of species feed by scraping algae off dead coral rock or rubble and sand. In the process much of the coral surface is crushed and digested. They produce vast quantities of sand this way which often is released during displays and swimming over reef-crests. As these fishes occur abundantly on coral reefs, often feeding in densely packed large schools, they are an important component in the reef building process. Most species have differently coloured stages as juvenile, and between sexes. The male usually has a mixture of brilliant contrasting colours, though green is usually the dominating colour.

Spawning is usually on dusk and most species group together in strategic places for the eggs to drift away with currents to open sea. Post-larvae settle when about 12-15 mm long.

Blue-barred parrotfish

L: To 75 cm. Di: Indo-Pacific, throughout our area.
De: 3-50 m. Ge: Coastal reefs, shipwrecks and harbours, often in silty conditions. Small juveniles in small groups and adults congregate in various numbers. ♀ right, ♂ below.

Scarus ghobban Forsskal, 1775

Ember parrotfish
L: To 55 cm. Di: Indo-Pacific.
De: 3-40 m. Ge: Juvenile with 2 somewhat indistinct dark spots on outer posterior scales on caudal fin and pale area heading dorsal fin. Juveniles solitair, adults in pairs, usually seen mixed with other parrotfish species to feed in school fashion. ♂ right, ♀ below.

Scarus rubroviolaceus (Bleeker, 1849)

PARROTFISHES • SCARIDAE

Three-colour parrotfish

L: To 40 cm. Di: Indo-Pacific, throughout our area.
De: 6-30 m. Gen: Clear coastal and outer reefs on crests and drop-offs. Usually in small aggregations of mixed sexes. ♀ with distinct colouration, but ♂ (below) among several similar species.

Scarus tricolor Bleeker, 1847

Forsten's parrotfish
L: To 45 cm. Di: West Pacific and east Indian Ocean.
De: 3-30 m. Ge: Clear coastal to outer reef crests, lagoons and drop-offs. Adults occurs in loose groups, seemingly solitair. ♀ has distinct colour pattern but ♂ (below) similar to other species. Juv. very dark with a small white side spot.

Scarus forsteni Bleeker, 1861

Orange-blotch parrotfish

Length: To 40 cm.
Distribution: West Pacific from Java along mainland Asia to tropical Japan.
Depth: 3-30 m.
General: The male is readily identified by the large brilliant orange patch on the sides. A common species in the Okinawa region where it attains about 30 cm, but specimens in Java grow much larger where it is uncommon and usually seem mixed with other schooling species. Photo shows ♂.

Scarus bowersi (Snyder, 1909)

PARROTFISHES • SCARIDAE

Yellow-tail parrotfish

Length: To 30 cm.
Distribution: Indonesia to southern Japan.
Depth: 3-30 m.
General: Coastal, often silty reef passages with mixed soft and hard corals. Juveniles and females are often common but males are usually seen solitair in those areas. Juveniles are distinct with a broadly stripes body and identified from other similar species by the black spot in the anal fin. Females have a bright yellow tail.

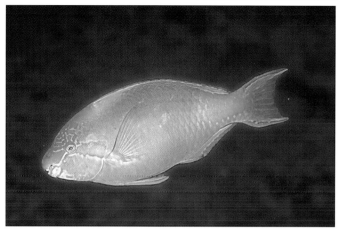

Scarus hypselopterus (Bleeker, 1853)

Dusky parrotfish

Length: To 35 cm.
Distribution: Indo-Pacific, throughout our area with two forms.
Depth: 3-40 m.
General: Pacific form of male shown. Indian Ocean form with green stripe instead of spot behind eye and young males are more striped which is more evident in females. Both forms are sympatric in Java. Coastal, often silty habitat, on crests and drop-offs and protected outer reef areas.

Scarus niger Forsskal, 1775

Yellow-head parrotfish

Length: To 30 cm.
Distribution: Widespread tropical west Pacific.
Depth: 3-20 m.
General: Males are easily recognised by their bright yellow head are are commonly seen on Indonesian reefs, though usually solitair. Dark brown females are usually nearby and in small groups feeding together on shallow reef crests. Clear coastal and protected outer reef habitats.

Scarus spinus Kner, 1868

237

PARROTFISHES • SCARIDAE

Scarus sp

Black-tail parrotfish

Length: To 25 cm.
Distribution: Indonesia, Philippines and Micronesia.
Depth: 15-40 m.
General: Unusual for a parrotfish, it seem to prefer deeper water. Clear coastal reef channels and along the base of drop-offs. Juveniles are distinct by their black tail and swim in small groups. Male usually seen solitair in the same areas, turning on a bright yellow patch above pectoral fin base during display.

Cetoscarus bicolor (Rüppell, 1829)

Two-colour parrotfish
L: To 90 cm. Di: Indo-Pacific, throughout our area.
De: 1-40 m. Ge: Juvenile shown in photograph is mostly noted by divers in the Pacific part of the range. Adults here are rather shy and look completely different. Juveniles are solitair and adults group, usually led by a very colourful male (below).

Bolbometopon muricatum (Cuvier & Valenciennes, 1840)

Humphead parrotfish
L: To 1.2 m. Di: Indo-Pacific.
De: 1-40 m. Ge: A large and impressive species. Outer reef slopes in schools reminiscent to a herd of Bison, grazing coral heads, scraping algeas on the sides. They sleep every night in the same crevices and may travel great distances during the day to feed.

SAND DIVERS • TRICHONOTIDAE

A small but interesting small family of slender fishes found widespread in the tropical Indo-Pacific. A single genus with about 4 species and probably several undescribed ones. They are strictly found on sand flats and slopes, and usually in tidal current zones where they occur in small to large aggregations. Most of the time they are buried in the sand and have just their eyes exposed, only when conditions for feeding are right the groups can be seen hoovering above the substrate, waiting for zooplankton coming past.

Only the most observent diver will see these fishes as they quickly dive into the sand when approached too quickly, or from the wrong direction. Males differ from the others by there extended anterior dorsal fin spines, which in some species are greatly produced. Females outnumber males in groups and are smaller. Males regularly display to females and to each other by raising the long rays of the dorsal fin and lowering the ventral fins. Nothing is known about spawning. Pelagic larvae are known to reach almost 20 mm before settling.

Long-rayed sand-diver

Length: To 18 cm (Japan, usually to 10 cm near equator).
Distribution: Throughout Indonesia and New Guinea to southern Japan.
Depth: 1-40 m.
General: A common schooling species, often just of the shore line on shallow sand flats but also on deep slopes at various depth levels depending on currents and type of sand. Highly variable in colour and some geographical variation, some of which were thought to be different species until its distribution was realized.

Trichonotus elegans Shimada & Yoshino, 1984

Blue-spotted sand-diver

Length: To 22 cm.
Distribution: Widespread tropical Indo-Pacific.
Depth: 3-40 m.
General: Clear coastal sand flats and slopes, often mixed with above species or forms small loose groups. Usually perched with ventral fins on sand, waiting for prey to come in range. Several geographical variations and perhaps comprises several species. Indonesian form shown which has usually three short elongate anterior dorsal spines.

Trichonotus setigerus Schneider, 1801

GRUBFISHES • PINGUIPEDIDIAE

A moderately large family comprising about 4 genera and probably over 60 species, but revision is badly needed. The distribution of most species in not known because of confusion with similar species. Small slender fishes also known as weevers or sand-smelts. Many occur in very deep water and are only known from trawls but they are also well represented in the shallows with many species in south-east Asian seas. The reef dwellers are typically found on sand or rubble patches where they perch themselves on the ventral fins. Very active during the day, quickly dashing about or under rocks for cover. Some species make burrows or use those made by other creatures. There is little change in colour during growth, and the differences between sexes are slight. Some species pair and often the differences between sexes is just the markings on the lower lips. Only a single genus *Parapercis* is recognised in the Indo-Pacific. They feed on a variety of small invertebrates and fishes, a few species have adapted to zooplankton and may swim high above the substrate. Most species are small, ranging from about 15 to 25 cm as adult.

Thousand-spot grubfish

L: To 18 cm Di: Indo-Pacific, throughout our area.
De: 2-15 m. Ge: Mixed coral-sand habitat, usually shallow flats but deeper down rubble slopes. Tail sometimes black and than similar to *P. hexophthalma* below.

Parapercis millepunctata (Günther, 1860)

False-eye grubfish

Length: To 24 cm
Distribution: West Pacific, eastern Indonsia to Philippines.
Depth: 2-15 m.
General: Coastal and outer reef crests and rubble slopes. Similar to Black-banded grubfish *P. tetracantha* throughout our area.

Parapercis clathrata Ogilby, 1911

GRUBFISHES • PINGUIPEDIDIAE

Peppered grubfish
Length: To 15 cm
Distribution: Indonesia, but probably more widespread.
Depth: 2-15 m.
General: Several similar species confused. Coastal rocky reefs among boulders on sand and slopes with rocks. Below a common but undescribed species below from our area.

Parapercis xanthozona (Bleeker, 1849)

Lyre-tail grubfish

Length: To 18 cm
Distribution: Indo-Pacific, throughout our area.
Depth: 10-50(+) m.
General: Coastal, current-prone slopes with coral or rock outcrops on sand, usually in small aggregations, being unusual by often feeding high above the substrate on zooplankton. Adults with extended caudal fin tips. Variable in general colour such as yellow on sides of head and pink to red banding or blotching.

Parapercis schauinslandi (Steindachner, 1900)

Sharp-nose grubfish

L: To 15 cm
Di: Indo-Pacific, throughout our area. De: 2-20 m.
Ge: Rocky reefs to seagrass beds with rubble patched. Enters large estuaries. Several similar undescribed ones, such as below the Three-line grubfish, widespread in our area.

Parapercis cylindrica (Bloch, 1797)

241

CONVICT BLENNIES • PHOLIDICHTHYIDAE

A single species family with restricted distribution in the west Pacific. They are goby-like but apparently closer related to the blennies. Like the blennies they lack scales, but unlike them lack the usually well developed teeth and instead have small conical teeth. They also lack fin spines and lateral line. During the day small juveniles occur in great numbers, sometimes clouding the water column along protected parts of drop-offs. Schools can be several metres long when swimming densely packed. They are often mistaken for coral- or striped catfishes by divers but it is unlikely that mimicry is the case. Swims closely packed when on the move or when threatened. At night they settle on the reefs and attach themselves with sticky threads secreted from large pores between the eyes. When about 10 cm the striped pattern begins breaking up into spots and they settle in pairs in the substrate. Adults are spotted and rarely seen (see photo below), living secretively in borrows which probably serve as nests as well. Babies have been seen emerging from these holes, so they lay eggs or are live-bearers. Easily kept in aquaria but ideally with other small species or in an invertebrate tank.

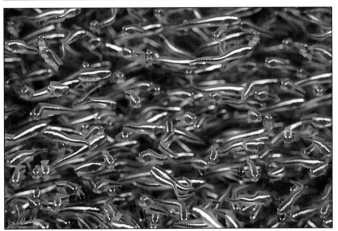

Convict blenny

General: Only small juveniles commonly observed. Sheltered coastal bays to deep outer reef walls. Nearly always in massive schools which spread out when feeding on zooplankton over a great distance or depth. Adults shown below.

Pholidichthys leucotaenia Bleeker, 1856

TRIPLEFINS • TRIPTERYGIIDAE

A very large family of mostly tiny fishes a few cm long. Probably more than 20 genera and about 200 species, most of which appear to be undescribed. A difficult family with many similar species living secretive in reefs and many are only collected when chemicals are used. They are related to the blennies but have scales and the dorsal fin is divided into three separate parts, hence their common name.

A few species sit exposed on reefs and can easily be mistaken for small gobies but have a more pointed snout and perch themselves on narrow ventral fins instead of the typical goby-disc. They are carnivorous, feeding mainly on small crustaceans. Males of some species turn-on bright colour to attract gravid females to its nesting site to lay her eggs. Hatchlings are about 4 mm long and swim to the surface and are pelagic. Postlarvae settle at about 10 mm length.

TRIPLEFINS • TRIPTERYGIIDAE

Striped triplefin

Length: To 40 mm.
Distribution: Widespread tropical Indo-Pacific, including all of our area.
Depth 6-30 m.
General: Clear coastal reefs, usually in small groups on encrusting sponges. This is one of the most colourful and most observed species by those taking notice of the little ones, easily identified by the thick red stripes over top and sides.

Helcogramma striata Hansen, 1986

Yellow-lip triplefin

Length: To 45 mm.
Distribution: Indo-Pacific, including all of our area.
Depth 3-20 m.
General: Coral rich reefs, usually on sides of large flat corals or encrusting sponges. Below a common but undescribed species from our area.

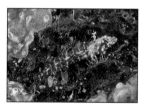

Helcogramma gymnauchen (Weber, 1909)

Highcrest triplefin

Length: To 25 mm.
Distribution: Indo-Pacific, including all of our area.
Depth 3-20 m.
General: Similar species in our area. Most have no easy mark for identification. Lives just below high energy zones on reefs. Unidentified male from Andaman Sea below.

Enneapterygius pusillus Rüppell, 1835

BLENNIES • BLENNIIDAE

A very large and complicated family of usually very small species, with more than 50 genera and well over 300 species, most of which distributed in tropical seas. They feature a slimy skin, rather than scales, and teeth consist of a comb-like arrangement in each jaw, with greatly enlarged canines in certain species. Presently various subfamilies and within these there are tribes. Some blennies mimic other other fishes, including other blennies. The best known is the copy of the cleaner wrasse. Primarily benthic fishes, but some specialised species feed on zoo-plankton or attack other fish well above the substrate feeding on their external parts, approaching with trickery, such as mimicing harmless species. The bottom dwelling species feed on a mixed diet of algae and invertebrates.

Most blennies take refuge in small holes in rocks or discarded worm tubes or shells. Males attract the gravid females by dancing and displaying their colours to deposit the eggs inside which is than guarded by the male. After a few weeks the tiny transparent hatchlings about two mm long swim to the surface, usually timed with the tides and taken out to sea by the current. Postlarvae settle when about 2 mm long.

Xiphasia setifer Swainson, 1839

Hairtail blenny

Length: To 40 cm.
Distribution: Widespread tropical Indo-Pacific, containing all of our area, ranging to sub-tropical zones.
Depth: 1-50 m.
General: Coastal sand and mud flats. Lives in vertical holes in mud made by other creatures. Usually just head exposed, occasionally swimming to another hole. A distinct species with eel-like body which typical for a blenny, backs into hole tail first.

Aspidontus dussumieri (Valenciennes, 1836)

Lance blenny

Length: To 12 cm.
Distribution: Widespread tropical Indo-Pacific, containing all of our area, ranging to sub-tropical zones.
Depth: 3-20 m.
General: Caudal fin lanceolate in adults, sometimes with long filaments centrally. Usually brownish or yellowish above and pale below, with a dark band from snout to caudal fin. Coastal, protected bays and sheltered reefs, usually staying close to the substrate, quickly retreating into empty tube-worm holes (backwards).

BLENNIES • BLENNIIDAE

Eyelash harptail blenny
L: To 10 cm. Di: Indonesia to Philippines to central Pacific. De: 5-20 m. Ge: Coastal reefs along low vertical vertical drops on coral rich shallow reefs. Long canine tooth on sides of lower jaws with venom gland at base, used for defence only. Mimic *Plagiotremus laudandus* below.

Meiacanthus atrodorsalis (Günther, 1877)

Smith's harptail blenny
L: To 75 mm. Di: Eastern Indian Ocean to Java. De: 1-20 m. Ge: Common in sheltered coastal bays. Juvenile *Scolopsis bilineata* mimics this species, lacking the yellow in other areas where it mimics other *Meiacanthus*. Photo right from Java, photo below (with spots) from the Andaman Sea.

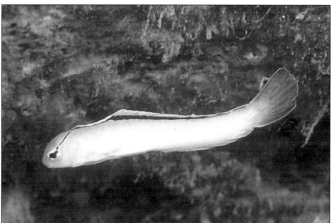

Meiacanthus smithi Klausewitz, 1961

Schooling harptail blenny

L: To 65 mm. Di: Eastern In-donesia to Philippines. De: 3-25 m. Ge: Clear coastal reefs in rich current prone areas where they form up to large schools to feed on zooplankton. *Meiacanthus abditus* Smith-Vaniz, 1987, from Indonesia, Philippines, below.

Meiacanthus ditrema Smith-Vaniz, 1976

BLENNIES • BLENNIIDAE

Shorthead sabretooth blenny

L: To 12 cm. Di: Indo-Pacific, containing all of our area.
De: 1-40 m. Ge: Yellowish on coral reefs where it mimics and swims with *Meiacanthus grammistes* (on left photo lower fish). It's large canines have no venom. Juvenile below.

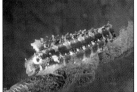

Petroscirtes breviceps (Valenciennes, 1836)

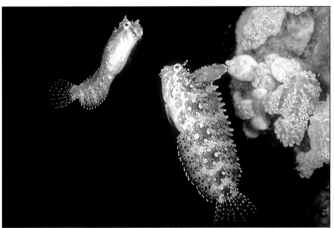

Crested sabretooth blenny

Length: To 75 mm.
Distribution: Widespread tropical Indo-Pacific, containing all of our area.
Depth: 1-10 m.
General: A common species on reefs with weeds or patches of seagrasses. Often with floating sargassum weeds and on jetty pilons in island lagoons where large algaes dominate. Sometimes floats along vertically by just using the pectoral fins. A distinct species by the elevated front part of the dorsal fin above the head.

Petroscirtes mitratus Rüppell, 1830

Xestus sabretooth blenny

Length: To 65 mm.
Distribution: Indo-Pacific, containing all of our area.
Depth: 1-20 m.
General: Sheltered coastal bays and lagoons on mixed sand and rubble reef. Juveniles hide under loose weeds, anemone mantels or underneath the upside-down jelly, *Cassiopea sp.* Adults use empty snail-like mollusc shells for home and nesting. Variable species, best recognised by the tufts under the chin.

Petroscirtes xestus Jordan & Seale, 1906

BLENNIES • BLENNIIDAE

Bath's combtooth blenny

L: To 40 mm. Di: Southern Island chain of Indonesia from Bali to the Banda Sea.
De: 3-30 m. Ge: Clear coastal and protected inner reef in rich invertebrate areas on reef crests and drop-offs to moderate depths. Male right, female below.

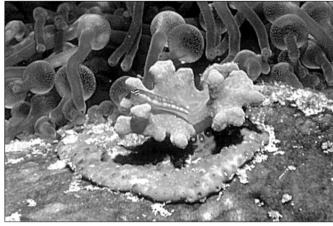

Ecsenius bathi Springer, 1988

White-lined combtooth blenny

Length: To 45 mm.
Distribution: Indonesia to Philippines and to Solomon Islands.
Depth: 15-40 m.
General: Mainly found on deep drop-offs along outer reefs. Singly or in pairs, usually resting on sponges, at moderate depths. A distinct species with the thin white lines, the upper-most breaks up with age, on a dark body.

Ecsenius pictus McKinney & Springer, 1976

Yellow-eyed combtooth blenny

Length: To 55 mm.
Distribution: Indonesia north to the Philippines.
Depth: 3-25 m.
General: Clear coastal reef crests and in coral rich lagoons on large bommies. Usually in small groups.
Shallow water form shown, in deep water the lower half of the body is usually very pale. Iris bright yellow, a small dark spot just behind eye, and black spot around anus.

Ecsenius melarchus McKinney & Springer, 1971

BLENNIES • BLENNIIDAE

Istiblennius edentulus (Forster, 1801)

Rippled rockskipper

Length: To 16 cm.
Distribution: Widespread tropical Indo-Pacific.
Depth: Intertidal.
General: Like other rockskippers they feed on algaes, scraped from rocks, when the tide is high just under the water line. When the water lever drops from wave action they hang-on out of the water. On low tide they can be found in rock pools and will jump about when disturbed.

Salarias segmentatus Bath & Randall, 1990

Red-spotted blenny

Length: To 11 cm.
Distribution: West Pacific, Indonesia, Flores to Micronesia.
Depth: 4-15 m.
General: Mainly in clear outer reef lagoons along base of rich coral slopes on coral bases with algae growth on which they graze. A distinct species in which the spots over back become bright red in adults. Occurs in small loose groups.

Crossosalarias macrospilus Smith-Vaniz & Springer, 1971

Big-spot blenny

Length: To 10 cm.
Distribution: Eastern Indonesia to Philippines and Samoa.
Depth: 1-10 m.
General: The large blotch, heading the dorsal fin identifies this species from other similar species. A solitary species found on hard coral and rock substrates with mixed coral and algae growth, the latter it grazes on. A common species but well camouflaged.

BLENNIES • BLENNIIDAE

Jeweled blenny

Length: To 14 cm.
Distribution: Widespread tropical Indo-Pacific, containing all of our area.
Depth: 0-6 m.
General: Algae rich reefs from intertidal to a few metres depth. Common in rubble reef which suffer occasional storm damage, feeding on algae building up on dead coral bases. Some geographical variations and differences between Indian and Pacific Ocean seem enough to regard them as separate species, both forms occur in Java seas. Right top Pacific form and Indian form below.

Salarias fasciatus (Bloch, 1786)

Leopard blenny
L: To 14 cm. Di: Widespread including South East Asia.
De: 1-15 m. Ge: Clear coastal reefs, and sheltered lagoons on coral substrates. Often common and usually shy, hiding in corals when approached. Some geographical colour variations from yellow to red. Eggs are laid on dead coral patches.

Exallias brevis (Kner, 1868)

DRAGONETS • CALLIONYMIDAE

A large family comprising at least 9 genera and about 125 species, with numerous small species in the tropics and some may represent new species. Fishes with broad spiny heads, instead of scales a tough slimy skin which is of bad taste, and with a strong odour, giving some species the common name of stinkfish. The mouth is greatly protrusible, extending out in a downward angle. They are benthic fishes, many of which buried in sand for most of the time, whilst the reef dwellers hug the substrate closely, virtually skipping along with fins touching at all times.

During spawning the pair (occasionally an extra male joines in) rise slowly from the substrate with ventral fins together. Eggs float and the larvae are plantonic. Adults range from very shallow depths to 400 m, depending on the genus or species, and mostly occur in sandy habitats. A few small species are typical reef dwellers which do not bury in the substrate.

Synchiropus ocellatus (Pallas, 1770)

Marbled dragonet

L: To 65 mm.
Di: Indonesia, Malaysia and Philippines to Japan.
De: Intertidal to at least 50 m.
Ge: First dorsal fin in males very tall. Shallow coral reef flats and rocky areas in sheltered coastal bays and inner reefs.

Synchiropus bartelsi Fricke, 1981

Bartels' dragonet

L: To 45 mm.
Di: Eastern Indonesia, Flores, to Philippines. De: 3-35 m.
Ge: At Flores fairly common on coastal reef crests amongst coral head on rubble patches. Has series of white spots along lower sides and adults with tiny blue ocelli on back.

DRAGONETS • CALLIONYMIDAE

Moyer's dragonet

L: To 75 mm.
Di: Indonesia and Philippines, to Japan and Australia. Possibly east Indian Ocean. De: 6-50 m.
Ge: Very similar to *S. ocellatus* but with red bars on anal fin. Rocky coastal reef flats and slopes with short algae and sparse coral growth.

Synchiropus moyeri Zaiser & Fricke, 1985

Mandarinfish

L: To 60 mm. Di: Indonesia, New Guinea and Philippines to Japan and Australia. De: 3-30 m.
Ge: A locally common but difficult to find fish. In dense coral heads. Coastal, often silty habitat, in drab looking corals or amongst open rubble with thin algaes.

Synchiropus splendidus (Herre, 1927)

Black-orange dragonet

Length: To 12 mm.
Distribution: Flores, Indonesia.
Depth: 4 m.
General: Only known from a single male (photograph) from Maumere, Flores, Indonesia. On sand with fine algae growth on which it was picking, probably for tiny crustacea. Colour like some nudibranchs (*Dermatobranchus*) of similar size and probably a case of mimicry.
As in this family the male is largest, this is one of the smallest known fish species.

Synchiropus kuiteri Fricke, 1992

DRAGONETS • CALLIONYMIDAE

Picture dragonet

Length: 70 mm.
Distribution: Philippines and a few Indonesian localities.
Depth: 1-6 m.
General: Mainly in Philippines but reported south to Komodo Islands in southern Indonesia. A sub-species in north-western Australia. Shallow coastal reefs.

Synchiropus picturatus (Peters, 1876)

Blue-spotted sand dragonet

Length: To 50 mm.
Distribution: Indonesia and Micronesia.
Depth: 2-10 m.
General: Coastal reef sand flats, usually in small groups comprising a single male and a number of females. Buries in the sand and comes out to feed during the day with changing tides.

Callionymus simplicicornis Valenciennes, 1837

Fingered dragonet

L: To 18 cm. Di: East Indian Ocean and W. Pacific, all of our area. De: 1-50 m. Ge: Sand and mud flats or slopes from very shallow depths to deep mud flats. Sometimes buries during the day and raises fins when disturbed before dashing off. Juvenile below.

Dactylopus dactylopus (Bennett, 1837)

252

GOBIES • GOBIIDAE

The largest family of marine fishes, comprising over 200 genera, and an estimated 1500 species world wide. The majority of species are tropical, and live on reefs or nearby on sand. Primarily benthic marine fishes, usually in touch with the substrate, a few readily hoover above. Many of the sand gobies bury in the sand at night, others use burrows. Some live in brackish water and enter lower reaches of streams, a few are totally adapted to fresh water. Because of their small size and secretive habits, many new species are being discovered on a regular basis, especially in the south-east Asian region, and the total number of species is even difficult to gues at this stage. Many species are highly specialised and associate with various invertebrates, ranging from certain coral species and sponges to specific crustaceans, especially alpheid shrimps.

Most gobies pair to breed and use burrows, narrow ledges, empty shells or any small enclosure fornesting site and produce several hundreds of sticky eggs which are laid in a neat patch. The male or both sexes guard the egg and nesting site. Hatching larvae are attracted by light and swim quickly to the surface.

Crab-eyed goby

Length: To 10 cm.
Distribution: West Pacific, to Philippines and to Solomon Islands.
Depth: 6-30 m.
General: Inhabits protected coastal sand areas along reefs and in deep fine-sand lagoons with rubble pieces. Pairs close together, taking turn feeding by taking mouths full of sand to filter for small prey. When alarmed it raises its dorsal fins and waves the ventral fin below and moves in rocking motion, in all reminiscent to a crab.

Signigobius biocellatus Hoese & Allen, 1976

Broad-barred sleepergoby

Length: To 12 cm.
Distribution: Indo-Pacific, throughout our area.
Depth: 10-30 m.
General: A distinct but rare species, known from less than 30 specimens, but ranging from east Africa to eastern Papua New Guinea to Australia and Japan. Coastal sand slopes on fine sand or silt substrate. Usually found single and shy, easier approached as a pair.

Valenciennea wardii (Playfair, 1867)

GOBIES • GOBIIDAE

Valenciennea randalli Hoese & Larson, 1994

Green-band sleepergoby

Length: To 16 cm.
Distribution: From Java, Indonesia east throughout our area.
Depth: 10-25 m.
General: Coastal silty mud or fine sand zones along the bottom of reef slopes. Adults in pairs and juveniles in small groups or pairs. The green band on the cheek identifies this species in the south-east Asian region.

Valenciennea puellaris (Tomiyama, 1956)

Orange-spotted sleepergoby

Length: To 14 cm.
Distribution: Throughout our area, several geographical forms.
Depth: 3-25 m.
General: A distinct species, pale with large orange spots, vertical barred in some areas. Coastal reef slopes and inner reefs, usually in rubble zones, juveniles in small aggregations, adults in pairs. Ranges to warm-temperate zone where it may grow much larger.

Amblygobius nocturnus (Herre, 1945)

Pyjamagoby

Length: To 10 cm.
Distribution: Widespread throughout our area.
Depth: 1-30 m.
General: Coastal silty lagoons and rubble-sand along reef edges. Young in small groups and adults in pairs, behaving identical to sleepergobies and also in looks they resemble these other gobies.
Digs burrows and feeds by filtering mouth-fulls of sand.

GOBIES • GOBIIDAE

Pink-lined goby
L: To 10 cm. Di: Widespread throughout our area. De: 3-30 m. Ge: Coastal and inner reefs, lagoons and slopes. Course rubble and low reef on sand zones. Juveniles singly, adults in pairs and usually in depths of about 10 m. Very similar to A. buanensis, an Indian Ocean species ranging to Java (below).

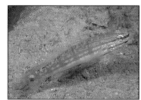

Amblygobius decussatus (Bleeker, 1855)

Byno goby

Length: To 10 cm.
Distribution: Indonesia, Java, to northern Australia.
Depth: 1-10 m.
General: Clear inshore lagoons and protected bays on fine sand with rubble, often on very shallow flats just below intertidal zone. Make borrows under solid objects and adults occur in pairs, feeding by filtering mouth-fulls of sand.

Amblygobius bynoensis (Richardson, 1844)

Sphynx goby

L: To 10 cm. Di: West Pacific. De: 1-10 m. Ge: Variable in relation to habitat: greenish in seagrasses and brownish on rubble-algae reef in still lagoons. Very shallow on small sand patches with weeds and seagrasses. The similar A. phalaena lives on reefs (below).

Amblygobius sphynx (Valenciennes, 1837)

GOBIES • GOBIIDAE

Hector's goby
L: To 85 mm. Di: Tropical Indian Ocean, along Asian mainland to southern Japan. De: 3-30 m. Ge: Clear coastal reefs and protected bays with rich coral and algae growth, slopes and shallow drop-offs, areas subject to moderate currents. Similar to A. rainfordi (below) from W. Pacific.

Amblygobius hectori (Smith, 1956)

Black-ray shrimpgoby

Length: To 6 cm.
Distribution: Widespread Indonesia, Malaysia and Philippines.
Depth: 8-40 m.
General: Coastal sand flats and slopes with sparse rubble, current prone areas.
Adults nearly always in pairs, hoovering just above burrow when feeding time, waiting for zooplankton to come in reach. Lives with Alpheus randalli who constructs the burrow.

Stonogobiops nematodes Hoese & Randall, 1982

Blue shrimpgoby
L: To 65 mm. Di: Indonesia, Java, Bali and Flores. D: 10-30 m. Ge: Coastal mud and sand slopes. A little known pairing species, apparently undescribed, which looks metalic blue with natural light. Lives with a brown-banded alpheid shrimp. The rare M. nigrivirgata, in Bali below.

Myersina sp.

GOBIES • GOBIIDAE

Smiling shrimpgoby

Length: To 8 cm.
Distribution: Indo-Pacific, throughout our area.
Depth: 6-30 m.
General: A common variable species from bright yellow to brown or black. First dorsal fin elongated in males.
Lives singly or in pairs in protected coastal bays on mixed sand and mud substrates with a grey-brown alpheid shrimp.

Mahidolia mystacina (Valenciennes, 1837)

Sailfin shrimpgoby

Length: To 12 cm.
Distribution: Indonesia to southern Japan, Solomon Islands and to northern Great Barrier Reef.
Depth: 10-50 m.
General: Usually found on white sand patches in caves or below large overhangs along drop-offs on inner and outer reefs. Singly or in pairs, often several pairs in large caves. Lives with a pale brownish, pale banded alpheid shrimp. One of the most distinctive species with the large sailfin-like first dorsal fin and striking colouration.

Amblyeleotris randalli Hoese & Steene, 1978

257

GOBIES • GOBIIDAE

Amblyeleotris latifasciata Polunin & Lubbock, 1979

Metalic shrimpgoby

Length: To 13 cm.
Distribution: Indonesia.
Depth: 10-40 m.
General: Coastal sand slopes with sparse rubble at moderate depths of about 15 m.
Pale with little colour on white sand but darker and colourful on dark sand.
Usually observed solitair with one or two marbled brown and white alpheid shrimps.

Amblyeleotris gymnocephala (Bleeker, 1853)

Red-margin shrimpgoby

Length: To 12 cm.
Distribution: Widespread in Indonesia, probably in adjacent areas.
Depth: 6-40 m.
General: A thin red outer margin of the whitish median fins identifies this species. Coastal white to dark sand and mud slopes, singly or in pairs with a marbled brown and white alpheid shrimp.

Amblyeleotris sp

Flag-tail shrimpgoby

Length: To 13 cm.
Distribution: Indonesia, Bali to Flores, but probable widespread.
Depth: 3-30 m.
General: Coastal sand flats, slopes and in deep lagoons. Distinct tail identifies this species, body colour varies with colour of sand.
Usually singly, living with the red-banded alpheid shrimp *Alpheus randalli*.

GOBIES • GOBIIDAE

Lagoon shrimpgoby

Length: To 12 cm.
Distribution: Indonesia, Java to Flores.
Depth: 1-6 m.
General: Well protected lagoons surrounded by inter-tidal reef flats, from silty coastal to clear outer reefs in very shallow depths.
Lives in pairs with black alpheid shrimp, and usually numerous pairs present in relatively small areas.

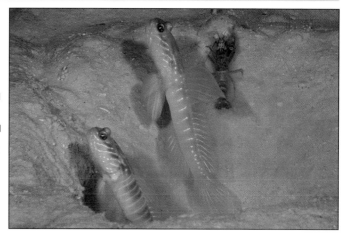

Cryptocentrus cyanotaenia (Bleeker, 1853)

Singapore shrimpgoby

Length: To 10 cm.
Distribution: Widespread Indonesia, Singapore, Malaysia to southern Japan and northern Australia.
Depth: 1-10 m.
General: Still coastal sand slopes and flats, often just beyont inter-tidal zone.
Singly or in pairs, living with dusky banded alpheid shrimp. Best recognised by the diagonal body bands and blue edged red spots on head.

Cryptocentrus singapurensis (Herre, 1936)

Spot-face shrimpgoby

Length: To 15 cm.
Distribution: Indonesia, Java and Bali but probably more widespread.
Depth: 10-40 m.
General: Deep coastal mud slopes, often well away from burrow and seems less reliant of alpheid shrimps.
Large adults seen singly resting on mud, juveniles usually at the entrance of burrows and living with alpheid shrimps.

Cryptocentrus polyophthalmus (Bleeker, 1853)

GOBIES • GOBIIDAE

Big mudgoby

Length: To 20 cm.
Distribution: Widespread Indonesia, Singapore and New Guinea.
Depth: 10-50 m.
General: Coastal mud flats near the base of slopes. Conspiciously sits on top of the mud but quickly dives into the soft substrate when approached too close. Occurs solitair, but usually in small numbers spread over an area.

Oxyurichthys ophthalmolepis Bleeker, 1856

Poisonous goby

Length: To 15 cm.
Distribution: Widespread tropical Indo-Pacific from east Africa throughout the west Pacific.
Depth: 6-30 m.
General: Silty coastal sand and mud bays, singly or in small loose groups, occasionally in pairs. Contains a toxin in the skin to deter predators, and for this reason it is easily approached at short range for getting good photographs.

Yongeichthys nebulosus (Forsskal, 1775)

Blue-speckled rubble goby

Length: To 35 mm.
Distribution: Andaman sea, Indonesia to New Guinea and north to southern Japan.
Depth: 6-40 m.
General: A common species on bare rubble slopes, usually in large spread out numbers, looking black at distance, hoovering just above the substrate to feed on zooplankton. Coastal to inner reefs in current prone areas.

Asterropteryx ensifera (Bleeker, 1874)

GOBIES • GOBIIDAE

Puntang goby

Length: To 14 cm.
Distribution: West Pacific, from Adaman Sea to Fiji and from Australia to Japan, including all of our area.
Depth: 1-10 m.
General: Shallow coastal lagoons and rock pools, often in freshwater run-offs. Sits out in the open on rocks or coral rubble, usually seen singly. This species is best recognised by the spotted pattern in the first dorsal fins.

Exyrias puntang (Bleeker, 1851)

Mud-reef goby

Length: To 15 cm.
Distribution: Widespread tropical Indo-Pacific, from east Africa to southern Japan and Australia, including all of our area.
Depth: 3-20 m.
General: Protected coastal to outer reefs on silty coral slopes, away from currents, usually along base of staghorn corals. Numerous in some areas but spread out as solitair individuals, regularily feeds by taking large mouths full of sand which are filtered for microorganisms by the gill rakers.

Exyrias belissimus (Smith, 1959)

Filamented sand goby

Length: To 12 cm.
Distribution: Widespread Indonesia to Philippines and Micronesia.
Depth: 5-20 m.
General: A distinct species with the long filaments on the first dorsal fin. Was thought to be the juvenile of the above Mud-reef goby but at Java's north coast both species are common and obviously are good species.
The photograph shows a large adult about 12 cm long.

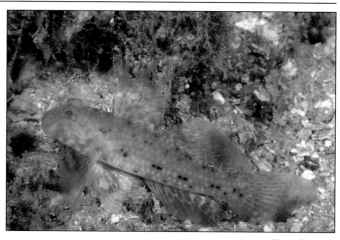

Exyrias sp

GOBIES • GOBIIDAE

Yellow coral goby
L: To 30 mm. Di: Widespread, from Java east in our area. De: 1-10 m. Ge: One of many coral goby which mostly live secretively amongst branches of Acropora corals. This one is the most obvious because of the yellow colour. Another yellow goby *Lubrigobius exiquus* lives on sand (below).

Gobiodon okinawae Sawada, Arai & Abe, 1972

Ring-eye pygmy goby
Length: To 40 mm.
Distribution: Indonesia, probably widespread.
Depth: 6-50 m.
General: A common species throughout Indonesia, found resting on sponges along rich drop-off and slopes from coastal to outer reefs. Distinct in colouration and appears to be undescribed.

Trimma sp

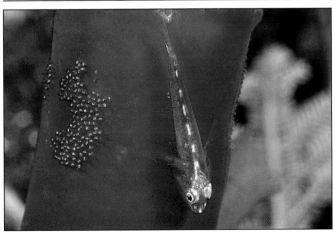

Many-host goby
Length: To 26 mm.
Distribution: Widespread tropical Indo-Pacific, including all of our area.
Depth: 2-20 m.
General: This species is the least choosy for a host, it is known to live on soft corals, stony corals, sponges, ascidians. molluscs, sea-cucumbers and algae. Eggs are deposited near or direct on host and guarded by the male.
Found in protected coastal bays to inner reef lagoons. The specimen in the photographs, left and page 263, Bali, Indonesia.

Pleurosicya mossambica Smith, 1959

GOBIES • GOBIIDAE

Pleurosicya boldinghi Weber, 1913

Soft-coral goby

Length: To 45 mm.
Distribution: Indo-Pacific from Africa to New Guinea and Australia to Japan.
Depth: 20-70+ m.
General: A stocky white looking goby found on soft corals, especially large *Dendronephthya spp* and usually in small groups.
Deep coastal slopes with isolated soft coral colonies and usually in depth over 25 m, one specimen known from 130 m trawl.

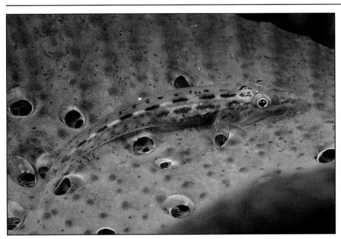

Pleurosicya elongata Larson, 1990

Slender sponge goby

Length: To 40 mm.
Distribution: Indonesia, New Guinea and northern Australia, probably widespread.
Depth: 3-50 m.
General: A distinct species which commonly lives on fan sponges, and mainly found in small groups on the underside. The top of the sponge is drab greenish-brown but the underside purple to which the goby has adapted.
The floppy sponges can be turned over and the fish is easily photographed.

Bryaninops loki Larson, 1985

Loki whip-goby

Length: To 30 mm.
Distribution: Widespread tropical Indo-Pacific, various locations ranging to warm temperate zones in Australia and Japan, common Indonesia and Philippines.
Depth: 3-30 m.
General: Common but camouflaged and sneaky, hiding on blind side of host. Coastal waters, often silty, on smooth whips and gorgonians.
Found in pairs or small groups, depending on size of host. Juveniles are completely transparent.

DART GOBIES • MICRODESMIDAE

A moderately large family of gobioid fishes, comprising about 12 genera and an estimated 45 species placed in 2 subfamilies: Microdesminae (Wormfishes) and Ptereleotrinae (Dartfishes). As the name suggests, the wormfishes are very long and slender, and the dartfishes are mostly small torpedo shaped slender fishes, They feature large eyes, and prominent dorsal and anal fins. Unlike most gobies which have a single cub-shaped pelvic fin, there are two separate and prominent pelvic fins. Some species are very colourfull and popular in the aquarium trade.

The dart gobies are benthic fishes, but hovering freely above the substrate, feeding primarily on zoo-plankton, either schooling or in pairs. They use burrows for refuge, holes in rocks or reef and sand dug away from below solid objects. The solitary or pairing species commonly use burrows made by other fishes and sometimes forcing themselves as unwanted guests. A few species may form large and dense schools to feed above the substrate which with the first sign of danger may all try to get into a single hole at once, creating the occasional problem. When currents are running they hover mostly in a stationary place above the substrate but quickly dart into a hole when worried by something, hence their common name.

Tiny and numerous eggs are laid in the burrows and hatchlings are pelagic up to 20-30 mm long before settling on the substrate, some species straight away forming schools. Being mostly transparent at first they can be difficult to identify and some species could only be identified when grown in an aquarium. All species are excellent aquarium fish and as planktivores are ideal to live in piece with invertebrates orientated set-ups.

The *Nemateleotris* species are very popular because of their attractive looks and small size. They usually do well in pairs. Some of the *Ptereleotris* species are also very beautiful and many can be kept in small groups. The slender worm-gobi es and ribbon-gobies are difficult to get and are not well known in hte aquarium trade, but should make interesting pets with similar requirements as the better known species. Despite their small size some of the species have lived for 7 or more years in captivity.

Lyre-tail dart goby

Length: To 12 cm.
Distribution: West Pacific, Indonesia, Philippines to New Guinea and warm-temperate zones in Japan and Australia.
Depth: 3-50 m.
General: Variable, greenish blue to bright blue, with pinkish anal fin. Continuous dorsal fin and lunate tail are diagnostic features.
Coastal to clear oceanic drop-offs in 3-50 m, areas prone to tidal currents.
Usually schooling in small to large numbers, sometimes found mixed with other similar species.

Ptereleotris monoptera Randall and Hoese, 1985

DART GOBIES • MICRODESMIDAE

Ptereleotris evides (Jordan and Hubbs, 1925)

Scissortail dart goby

Length: To 12 cm.
Distribution: Indian Ocean and West Pacific, including Andaman Sea, Singapore, Malaysia, Indonesia, Philippines to New Guinea and warm-temperate zones in Japan and Australia.
Depth: 3-50 m.
General: A distinct species with the large median fins and colouration: variable pale anteriorly, blue-grey posteriorly. Protected bays and lagoons, including on outer reefs. Usually in pairs swimming just above the bottom. Current prone areas, feeding on zooplankton.

Ptereleotris heteroptera (Bleeker, 1855)

Tail-spot dart goby

Length: To 10 cm.
Distribution: Widespread tropical Indo-Pacific, including all of our area.
Depth: 3-50 m.
General: Variable in colour with habitat, on light coloured sand pale sky-blue, juveniles often bright blue, with a distinct black blotch on caudal fin, the latter sometimes yellow. Shallow reef flats to deep reefs, usually on sand patches or fringing reefs to about 50 m. Mostly in pairs or sometimes in small aggregations.

Nemateleotris magnifica Fowler, 1938

Red fire goby

Length: To 75 mm.
Distribution: East Africa to Indonesia, Philippines to New Guinea and to southern Japan and northern Australia.
Depth: 6-50 m.
General: Easily identified, pale anteriorly, gradually becoming bright red posteriorly, from below 2nd dorsal fin on. Mostly in pairs or small aggregations along reef edges below drop-offs, sometimes in deep lagoons, but generally where tidal currents effect the area. An ideal and sought after species for the invertebrate aquarium.

DART GOBIES • MICRODESMIDAE

Purple fire goby

Length: To 75 mm.
Distribution: Mauritius to Indonesia, Philippines, southern Japan and northern Australia.
Depth: 20-70 m.
General: A deep water species, showing little colour at depth. Only in some places in Indonesia (Bali) in depths less than 30 m, usually seen at depths of about 50 m, at the base of deep drop-offs on rubble slopes. Nearly always in pairs, hoovering just above their burrow and quickly dives into borrow for cover when approached.

Nemateleotris decora Randall & Allen, 1973

Blue-barred ribbon goby

Length: To 20 cm.
Distribution: Indonesia, Philippines and Malaysia, probably Papua New Guinea.
Depth: 15-60 m.
General: One of several species which typically live on coastal and usually turbid areas. At Flores this species occurs in large numbers in often clear waters on the mud slopes in depths of about 25 m, in pairs or numerous neighbouring pairs which at distance look like a school when feeding just above the substrate on zooplankton in the currents. A truly magnificent sight as their blue colouration is bright at depth.

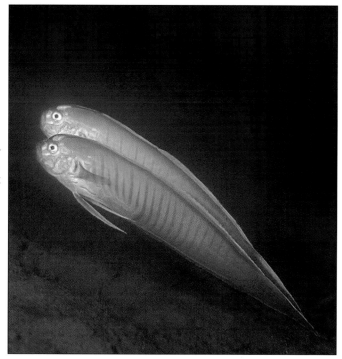

Oxymetopon cyanoctenosum Klausewitz & Condé, 1981

DART GOBIES • MICRODESMIDAE

Gunnellichthys viridescens Dawson, 1968

Orange-line worm goby

Length: To 12 cm.
Distribution: Indo-Pacific, throughout our area.
Depth: 10-50 m.
General: Coastal and inner reefs on sand-rubble flats and slopes. Solitair but sometimes locally common: Tulamben, Bali, a regularly dived area for many years, this species occurred commonly one year during May, but not seen there since, including the two following years about the same time of the year.

Gunnellichthys pleurotaenia Bleeker, 1858

Black-stripe worm goby

Length: To 12 cm.
Distribution: West Pacific, Indonesia, Java to Philippines, New Guinea and northern Australia.
Depth: 1-10 m.
General: A shallow coastal species, usually in the vicinity of seagrasses and mangroves, swimming along the edges over sand, readily to dive in any hole for cover. Solitair and only seen when snorkeling along seagrass beds or when using scuba in very shallow coastal lagoons.

Gunnelichthys monostigma Smith, 1958

Black-spot worm goby

Length: To 15 cm.
Distribution: Indo-Pacific, throughout our area.
Depth: 6-30 m.
General: Sheltered bays and lagoons on sand slopes. Hovers solitary above burrows of sand gobies.

RABBITFISHES • SIGANIDAE

Primarily a tropical family, comprising 2 subgenera and about 30 species, broadly distributed in the Indo-Pacific. All species have identical fin counts, except pectoral fins, which feature many large and sharp venomous spines. A sting is extremely painfull and pain lasts for about a half hour. Applying heat, eg. emerging the wound in very hot water gives instant relief, destroying the venom. The family is unique in having a spine at each end of the ventral fins with 3 rays in between. The species with a notably longer snout belong in the subgenus Lo. In the daytime most rabbitfishes are easily recognised by distinct colour patterns, especially those associating with coral reefs. Several species live in seagrass or algae areas and are very similar. Some work with these report spawning in synch with the lunar cycle, producing up to half million eggs per season which lasts about 3-4 months of the year. Larvae span is to about 4 weeks and postlarvae are only about 10 mm long settling in dense seagrasses or those associating with corals can be found amongst dense *Acropora* heads. They are mostly herbivorous.

Gold-spotted rabbitfish
L: To 40 cm. Di: Widespread tropical west Pacific and east Indian Ocean to Andaman Sea, including all of our area.
De: 3-45 m. Ge: Juveniles on shallow reef flats, including brackish water, and adults pair in deeper coastal waters on mixed sand and rocky substrates. Juvenile below.

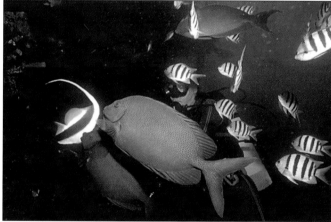

Siganus punctatus (Schneider, 1801)

Starry rabbitfish

Length: To 40 cm.
Distribution: Widespread tropical Indian Ocean, ranging east to Sunda Strait, Java.
Depth: 3-45 m.
General: Very similar to it's Pacific sibling species above and the two species overlap between the Andaman Sea and Java. It looks a lighter colour underwater and the spots are brown, rather than orange. The Red Sea population has a yellow tail and is regarded as a subspecies, however by comparison to other similar species they seem good species in their own right.

Siganus stellatus Forsskal, 1775

RABBITFISHES • SIGANIDAE

Siganus javus (Linnaeus, 1766)

Java rabbitfish

Length: To 40 cm.
Distribution: Widespread tropical Indo-Pacific, including all of our area.
Depth: 1-25 m.
General: Mainly coastal habitats from shallow estuarine lagoons as juveniles or adults, to adjacent coastal reefs where they are sometimes seen in groups feeding midwater on plankton.
Easily identified by the white spots or lines and the large black blotch in the caudal fin.

Siganus vermiculatus (Valenciennes, 1835)

Maze rabbitfish
L: To 50 cm. Di: Widespread tropical west Pacific and east Indian Ocean to Andaman Sea and India, including all of our area. De: 1-45 m. Ge: A highly variable species found from river mouths to deep and clear coastal reefs. Juveniles in brackish water. Highly variable in scribbled pattern, often breaking up into spots over most of the head. Adults singly or in large schools. A school at northern Sulawesi comprising adults showed a great range of patterns from fully vermiculated to mostly spotted patterns.
In the Seribu Island where this species is very common the juv. also showed great variation.

Siganus guttatus (Bloch, 1787)

Gold-saddle rabbitfish

Length: To 40 cm.
Distribution: Widespread tropical west Pacific and east Indian Ocean to Andaman Sea, including all of our area.
Depth: 3-35 m.
General: Mainly restricted to coastal rocky reefs with soft coral growth. Forms small groups, often seen hoovering in one place in the shelter of reefs. Its sibling species, *Siganus lineatus* is lined instead of spotted and occurs from the Philippines along eastern margin of our area to Australia.

RABBITFISHES • SIGANIDAE

Andaman foxface

Length: To 24 cm.
Distribution: Andaman Sea to Java, Indonesia.
Depth: 3-25 m.
General: Adults pair on clear coastal reefs. Juveniles inshore in rich coral growth.

Siganus magnificus (Burgess, 1977)

Black-blotch foxface

Length: To 25 cm.
Distribution: Tropical northwest Pacific, Philippines and Japan, and (?) north-western Australia.
Depth: 3-45 m.
General: Protected coastal waters. Juveniles from large schools in the shallows and graze on algaes and other weeds. Adults may form pairs or alos can occur in great schools over reefs.
A small population in Australia is virtually identical in colour and regarded as the same species. The similar yellow foxface *S. vulpinus* from our area south of Philippines below.

Siganus unimaculatus (Evermann & Seale, 1907)

RABBITFISHES • SIGANIDAE

Coral rabbitfish

Length: To 30 cm.
Distribution: Widespread tropical west Pacific and Indian Ocean to Seychelles, including all of our area.
Depth: 1-25 m.
General: Juveniles in *Acropora* corals, often in small numbers. Adults usually in pairs in rich invertebrate zones on coastal reefs and outer reef lagoons, grazing on algaes on the dead bases of corals.
The similar *S. puelloides* occurs in our area from Java west.

Siganus corallinus (Valenciennes, 1835)

Masked rabbitfish

Length: To 30 cm.
Distribution: Widespread tropical west Pacific and east Indian Ocean to Sumatra, probably including all of our area.
Depth: 3-35 m.
General: Rich invertebrate reefs.
Juveniles amongst branches of *Acropora* coral head, in small numbers or mixed with other species. With increasing size they start grazing on nearby algaes on the dead bases of corals.
Adults in pairs, usually secretive in staghorn corals in coral rich lagoons.

Siganus puellus (Schlegel, 1852)

Double-barred rabbitfish

Length: To 30 cm.
Distribution: Widespread tropical west Pacific and east Indian Ocean to India, including all of our area. Depth: 2-25 m.
General: Coastal habitats. Small juveniles were collected in pure fresh water at Flores. Adults in pairs or small aggregations on coastal, often turbid reefs, grazing on sand and rock substrate on shallow flats or crests of rocky reef. Replaced by sibling species, *S. doliatus,* in eastern-most part of our area, the latter ranging to central Pacific. Very similar, but has numerous blue vertical lines on the body.

Siganus virgatus (Valenciennes, 1835)

MOORISH IDOLS • ZANCLIDAE

This family is represented by a single widspread Indo-Pacific species. It shares many characters with the closely related surgeonfishes, Acanthuridae, and are placed together in the suborder Acanthuroidei, including the Siganidae. They share strongly compressed bodies, but in the moorish idol it is almost circular shaped. The snout is produced with a small mouth and teeth are long, slender and bristle-like, which are covered mostly by the fleshy lips. Adults develop sturdy spines in front of the eyes but lack spines or scutes on sides of caudal peduncle, which are a character in surgeonfishes, and lack venomous spines like in the rabbitfishes. The moorish idol is typified by the long white wimple like filament in the dorsalfin.

Eggs of the moorish idols are pelagic and the transparent larvae is pelagic for a long time, floating over a great distance, often into warm temperate zones and well beyond breeding areas. Large larvae already have the shape of the adult, including the filament in the dorsal fin.

Moorish idol

Length: To 22 cm.
Distribution: Widespread tropical Indo-Pacific, including all of our area, ranging to subtropical waters.
Depth: 1-180 m.
General: Distinctly marked with broad white and black banding and long white dorsal fin extention.
Juveniles expatriate well beyond breeding range into cooler zones. Coastal to outer reefs, often in shallow boulder areas feeding on algaes, but also venture to deeper zones where sponges are grazed. Settling juveniles are quite large, about 60 mm or more. Adults occur in pairs or small aggregation, but occasionally school in great numbers in some areas.

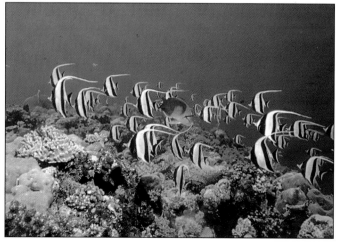

Zanclus cornutus (Linnaeus, 1758)

SURGEONFISHES • ACANTHURIDAE

A large circum-tropical family, comprising 3 subfamilies. The Acanthurinae largest, comprising 4 genera and about 50 species, featuring a single fixed spine on each side of the caudal peduncle, some of which are venomous. The spine is not hinged, as suggested in some literature, but when used in defense or when fighting, the tail is bent which makes the spine stick out. The Nasinae comprises about 15 species usually placed in a single genus, featuring 1 or 2 bony plates with spines on each side of the caudal peduncle. The more temperate Prionurinae comprises only a few species, featuring a series of bony peduncular plates.

Diet varies between species, some graze algaes, feed on zoo-plankton, or both, and others filter food from the substrate by digesting sand. The planktivorous school are often in great numbers, whilst benthic feeders either school or pair. The reef dwelling species are sometimes very colourful and juveniles may differ greatly from the adult in colour as well as shape. The spines on the tail are often surrounded by bright colours to serve as a warning, and is venomous in at least one species.

Blue-spine unicornfish

Length: To 70 cm.
Distribution: Widespread tropical Indo-Pacific, including all of our area.
Depth: 10-50+ m.
General: A common species in our area, adults usually in large numbers along deep drop-offs feeding on plankton in the current.
Males regularly displays and changes colours quickly to pale blue for a short time. Easily recognised by the blue tail which stands out even more in the natural light.

Naso hexacanthus (Bleeker, 1855)

Onespine unicornfish
L: To 45 cm. Di: Indo-Pacific, throughout our area.
De: 2-50+ m. Ge: Juveniles in large schools over open areas. Adults on slopes leading very deep, usually feeding in open water a long way from the reef. Returns to reef towards dusk, visiting cleaning stations and to take refuge for the night.

Naso thynnoides (Valenciennes, 1835)

SURGEONFISHES • ACANTHURIDAE

Humpnose unicornfish

Length: To 60 cm.
Distribution: Widespread tropical Indo-Pacific, including all of our area.
Depth: 2-20 m.
General: Large adults develop a large round hump above the mouth. Reef channels and protected coastal bays on rubble-algae flats and slopes. Sometimes congregates in large numbers in outer reef channels, normally solitair or in small numbers, usually staying at distance from divers.

Naso tuberosus Lacépède, 1802

Slender unicornfish

Length: To 60 cm.
Distribution: Widespread tropical west Pacific and east Indian Ocean, including all of our area.
Depth: 6-50+ m.
General: Always in small to moderate sized schools along slopes or drop-offs leading to very deep water. Clear coastal to outer reef habitats, feeding well away from the substrate on plankton but sleeps on the substrate by resting against rocks on sand.

Naso lopezi Herre, 1927

White-margin unicornfish

Length: To 1 m.
Distribution: Indo-Pacific, throughout our area.
Depth: 10-30+ m.
General: Large adults form large schools, usually along the edge of deep drop-offs, including in clear coastal waters. A shy species, best recognised by the white margin on the tail.

Naso annulatus (Quoy & Gaimard, 1825)

SURGEONFISHES • ACANTHURIDAE

Naso minor (Smith, 1966)

Little unicornfish
Length: To 25 cm.
Distribution: Widespread tropical Indo-Pacific, including all of our area.
Depth: 10-50 m.
General: Only recorded from Mozambique, Philippines and southern Japan, but occurs throughout our area. Swims in large schools along reef channels between inner reefs. The smallest of the unicorns, best recognised by the yellow tail and black spot surrounding the single peduncle spine. Usually misidentified as *N. thynnoides*, the next species which has a dark tail, no black caudal peduncle spot, and one less spine in the dorsal fin.

Naso unicornis (Forsskal, 1775)

Shortnose unicornfish

Length: To 70 cm.
Distribution: Widespread tropical Indo-Pacific, including all of our area.
Depth: 2-80 m.
General: Mostly grey to olivaceus, sometimes with broad grey anteriorly on sides. The blue caudal peduncle plates readily identifies this species, including juveniles.
Shallow reef flats to deep slopes, adults grazing algae or feeding in schools along drop-offs on plankton. Juveniles in sheltered coastal bays, feeding on benthic algaes.

Naso brevirostris (Valenciennes, 1835)

Longnose unicornfish

Length: To 50 cm.
Distribution: Widespread tropical Indo-Pacific, including all of our area.
Depth: 2-80 m.
General: Greenish-grey to brownish, variable with small spots or thin vertical lines developing in adults. Juveniles with white ring around caudal peducle, a feature shared by several other species.
Young on shallow reefs, feeding on benthic algae, adults usually along drop-offs feeding on zooplankton.

SURGEONFISHES • ACANTHURIDAE

Orange-spine unicornfish

Length: To 45 cm (excluding very long filaments on tail).
Distribution: Widespread tropical Indo-Pacific, including all of our area.
Depth: 1-50+ m.
General: Clear coastal algae rich rock reef, usually on shallow boulder or channel parts to graze fresh algae growth. Feeds singly, in pairs or sometimes in small aggregations and often joins large groups of other algae grazers including parrotfishes. Usually shallow, but reported to 90 m depth.

Naso lituratus (Forster, 1801)

Bignose unicornfish

Length: To 55 cm.
Distribution: Widespread tropical Indo-Pacific, including all of our area.
Depth: 2-50 m.
General: Clear coastal reefs adjacent to deep water, deep outer reef lagoons in current channels. Usually in loose schools in deep water, rising to shallow depth when feeding in currents on plankton. Sleep in caves at night in depths of 10 m or more. Only large adults develop the big nose and long filaments on the tail. Juveniles are finely spotted.

Naso vlamingii (Valenciennes, 1835)

SURGEONFISHES • ACANTHURIDAE

Palette surgeonfish

Length: To 30 cm.
Distribution: Widespread tropical Indo-Pacific, including all of our area.
Depth: 1-40 m.
General: The distinct colour readily identifies this species, Clear outer reefs, usually in schools feeding above the substrate on zooplankton, readily to dive among branching corals when approached. In Bali the adults feed mostly on benthic algae. Indonesian name is number 6 fish in relation the the black mark on the right side.

Paracanthurus hepatus (Linnaeus, 1766)

Orange-blotch surgeonfish
Length: To 35 cm.
Distribution: Indo-Pacific, throughout our area.
Depth: 2-50 m.
General: A benthic feeder, grazing algaes on rocks and sand among shallow reefs. Juveniles singly, often mixed with other juvenile surgeonfishes (see below).

Acanthurus olivaceus Forster, 1801

Powderblue surgeonfish

Length: To 23 cm.
Distribution: Widespread tropical Indian Ocean, ranging east to Bali.
Depth: 1-30 m.
General: Inshore surge zones to off-shore reef flats and slopes, often around rocky peaks breaking the surface. Juveniles singly in Indonesia and adults in small loose numbers. Some oceanic locations they form enormous schools.

Acanthurus leucosternon Bennett, 1832

SURGEONFISHES • ACANTHURIDAE

Pacific mimic surgeonfish

Length: To 20 cm.
Distribution: West Pacific, including Pacific part of our area, replaced by next species in Indian Ocean.
Depth: 2-40 m.
General: Juveniles mimic various *Centropyge* angelfishes depending on the area. Solitair in shallow coastal waters, grazing algae on rocks or sand. As mimic gains some freedom out in the open from predators, which have learned not to hunt the angelfishes because they cleverly and quickly move about, giving predators little chance to strike and look for easier prey. Adult shown below.

Acanthurus pyroferus Kittlitz, 1834

Indian mimic surgeonfish

Length: To 20 cm.
Distribution: Tropical Indian Ocean, from Sri Lanka to Bali.
Depth: 2-40 m.
General: Sibling species of *A. pyroferus*, above. Both species occur together in the Bali region of Indonesia. The right photo shows a mimic of the angelfish *Centropyge eibli*. The photo below an adult Indian mimic surgeonfish from the Andaman Sea.

Acanthurus tristis Tickell, 1888

SURGEONFISHES • ACANTHURIDAE

Velvet surgeonfish
L: To 20 cm. Di: West and east pacific. De: 1-50 m.
Ge: Inshore, shallow surge zones among large boulders and along cliff faces with cracks and ledges, but also on outer reef channels in current prone areas. Below the similar *A. japonicus* which occurs from Philippines to Japan.

Acanthurus nigricans Cuvier, 1829

Pale surgeonfish
L: To 45 cm. Di: Indo-Pacific, including all of our area.
De: 2-40 m. Ge: Adults feed mid water on zooplankton, juveniles feed on benthic algae, often in rocky estuaries occurring in small aggregations. Below variation from Andaman Sea.

Acanthurus mata Cuvier, 1829

Pencilled surgeonfish

Length: To 50 cm.
Distribution: Indo-Pacific, throughout our area but uncommon in eastern Indonesia.
Depth: 3-100 m.
General: Juveniles in large coastal estusries, grazing on algae covered rocks. Adults more oceanic in cave areas on sea mounts and islands.

Acanthurus dussumieri Valenciennes, 1835

SURGEONFISHES • ACANTHURIDAE

Eye-spot surgeonfish

Length: To 40 cm.
Distribution: Widespread tropical west Pacific and Indian Ocean to Maldives, including all of our area.
Depth: 10-50+ m.
General: Deep coastal to outer reef slopes and walls. Usually seen singly at moderate depths, grazing on the substrate. Large specimens are strongly convex on snout and become beautifully coloured. The small round spot immediately behind eye identifies this species.

Acanthurus bariene Lesson, 1830

Eye-line surgeonfish

Length: To 40 cm.
Distribution: Widespread tropical Indo-Pacific, including all of our area.
Depth: 3-40 m.
General: Caudal fin strongly lunate in adults. Clear coastal reefs and deep lagoons, usually seen solitair, grazing on small rock and boulder zones edging onto sandy areas. Adults best identified by the dark line behind eye.

Acanthurus nigricauda Duncker & Mohr, 1929

Yellow-fin surgeonfish

Length: To 56 cm.
Distribution: Widespread tropical Indo-Pacific, including all of our area.
Depth: 3-90 m.
General: Juveniles mainly coastal, often silty habitat. Adults prefer clear coastal slopes, lagoons and slopes along the base of outer reef drop-offs. A benthic feeder over mixed sand and rock substrate, adults usually in small groups and juveniles solitair. The largest species in the genus.

Acanthurus xanthopterus Valenciennes, 1835

SURGEONFISHES • ACANTHURIDAE

Acanthurus auranticavus Randall, 1956

Ringtail surgeonfish

Length: To 35 cm.
Distribution: Widespread tropical west Pacific and Indian Ocean to Maldives, including all of our area.
Depth: 3-50 m.
General: An indistinct species easily missed and mistaken for other similar species. Spot behind eye elongated and white band at tail base. When displaying an orange patch around tail spine. A common species on Bali's north coast, swimming in small groups or mixed with other species, grazing on rocks on the shallow flats.

Acanthurus maculiceps (Ahl, 1923)

Spot-face surgeonfish

Length: To 35 cm.
Distribution: Widespread tropical west Pacific and east Indian Ocean, including all of our area.
Depth: 1-20 m.
General: Shallow, clear coastal reef flats and slopes. Usually in small groups or mixed with other species to feed on algae, grazed from rocks. The spots on the face readily identifies this species. Body colour varies with mood from black to pale with fine lines.

Acanthurus fowleri de Beaufort, 1951

Horse-shoe surgeonfish

Length: To 45 cm.
Distribution: Indonesia, Philippines and Papua New Guinea.
Depth: 6-50+ m.
General: Clear coastal and outer reef habitat near deep water. Usually seen solitair deep down along drop-offs. Rarely seen in depths less than 20 m. At depth the blue snout, convex in large specimens, stands out, and in addition if close enough to see the blue horse-shoe shaped line behind eye, identification is made.

SURGEONFISHES • ACANTHURIDAE

Lined surgeonfish

Length: To 35 cm.
Distribution: Widespread tropical Indo-Pacific, including all of our area.
Depth: 0.5-10 m.
General: Caudal fin lunate, corners greatly extended in adults. An easily identified species by its colour pattern, even as small juvenile.
Usually in very shallow depths, schooling over reef crests with gutters, rarely deeper than 6 m. Juveniles shallow interdital zones, coastal protected boulder reef, and often in small aggregations.

Acanthurus lineatus (Linnaeus, 1758)

Convict surgeonfish

Length: To 26 cm.
Distribution: Widespread tropical Indo-Pacific, including all of our area.
Depth: 1-20 m.
General: Caudal fin slightly emarginate, corners not extended in adults.
A distinct species with little variation from juvenile to adult. Juveniles often in tidal pools. An algae grazer, which in some areas occur in massive numbers, completely overwelming aggresive damsels which often defend such areas in high spirits.

Acanthurus triostegus (Linnaeus, 1758)

Yellow-tip bristletooth

Length: To 16 cm.
Distribution: Indonesia, Philippines to Micronesia and Solomon Islands.
Depth: 10-50 m.
General: Clear coastal to outer reefs, on slopes and drop-offs with rich invertebrate growth. Juveniles singly, adults in small groups and mostly in depth of more than 20 m. A distinct species with the yellow tips on dorsal and anal fins.

Ctenochaetus tominiensis Randall, 1955

SURGEONFISHES • ACANTHURIDAE

Two-spot bristletooth

Length: To 20 cm.
Distribution: Indo-Pacific, including all of our area.
Depth: 2-50 m.
General: Algae zones on sheltered coastal reefs and in laggons on outer reefs. Juvenile (below) solitair, adults sometimes in small aggregations.

Ctenochaetus binotatus Randall, 1955

Goldring bristletooth

L: To 18 cm.
Di: Indo-Pacific, including all of our area. De: 2-30 m.
General: Coral rich reefs from coastal slopes to outer reef lagoons. Juvenile (below) highly variable from drab brown in some areas to bright yellow in others.

Ctenochaetus strigosus (Bennett, 1828)

Fine-lined bristletooth

Length: To 26 cm.
Distribution: Indo-Pacific, including all of our area.
Depth: 2-30 m.
General: Very common throughout our area on clear coastal to outer reef crests in small groups, forming large schools in more oceanic locations. Feeds on benthic algae, including those causing ciguatera poisoning if eaten.

Ctenochaetus striatus (Quoy & Gaimard, 1825)

SURGEONFISHES • ACANTHURIDAE

Pacific sailfin tang
L: To 40 cm. Di: Widespread tropical west Pacific to Hawaii and west to Java, Indonesia. De: 2-30 m. Ge: Range overlaps with Indian sibling species *Zebrasoma desjardinii* (Bennett, 1835), photo below, in Java. Juveniles in rich coral and algae reef in coastal and lagoon areas.

Zebrasoma veliferum (Bloch, 1797)

Brown tang

Length: To 20 cm.
Distribution: Widespread tropical Indo-Pacific, including all of our area.
Depth: 1-60 m.
General: Juvenile (below) secretive in rich coral growth in coastal and lagoon habitat. Adults form groups, raoming the reef flats for feeding sessions. A very attractive colour varation is shown here documented from the Philippines, right below.

Zebrasoma scopas (Cuvier, 1829)

285

BARRACUDAS • SPHYRAENIDAE

Barracudas are primary of tropical seas, comprising 1 genus and about 20 species globally distributed. Some species are large, reaching more than a metre in length and the great barracuda attains almost 2 m. They are fears predators, feeding mainly other fishes and squid, hunting alone or in groups, and some of the smaller species in close packed schools. Their bodies are streamlined and the large mouth is equipt with powerful jaws and sabre-like teeth. Most of the hunting is done during dawn and dusk, though they appear to patrol the reefs on the look-out to take advantage of any opportunity coming along for an easy feed.

Identification underwater is difficult for most species, especially the smaller ones as most can turn on or off lines and have different colours when swimming in schools midwater. Only a few have colour features to make a positive identification, the others are plain, mostly silvery and differ in position of fins and scale counts. Eggs are round and small, about 1 mm in diameter, and pelagic. Larvac hatch at about 2 mm and young settle in coastal sheltered waters when about 25 mm long. Some can be found in brackish water and in the mouth of rivers.

Great barracuda
L: To 1.9 m. Di: Indo-Pacific, throughout our area.
De: 0.5-100 m. Ge: Large specimens usually seen solitair on reefs and easily identified. Sometimes in small groups in deep water. Juveniles greenish dusky blotches in shallow coastal, brackish water. Pelagic colour below.

Sphyraena barracuda (Walbaum, 1792)

Chevron barracuda

Length: To 90 cm.
Distribution: Indo-Pacific, throughout our area.
Depth: 10-100 m.
General: A schooling species, juveniles coastal and adults in clear oceanic waters. Best identified by black upper and lower caudal fin margin and series of chevron bars in adults. Caudal fin in adults look black underwater. Similar to next species, *S. jello* with yellow tail and has shorter bars.

Sphyraena putnamiae Jordan & Seale, 1905

BARRACUDAS • SPHYRAENIDAE

Pickhandle barracuda

Length: To 1.5 m.
Distribution: Indo-Pacific, throughout our area.
Depth: 1-100 m.
General: A large species, schooling as juveniles but solitair or in small aggregations as adult. Bars on body saddle like, extending only just over lateral line and tail yellow in large individuals on reefs. Juveniles in coastal bays, adults on outer reefs but usually deep.

Sphyraena jello Cuvier, 1829

Blue lined barracuda

Length: To 1.5 m.
Distribution: Indo-Pacific, throughout our area.
Depth: 10-50 m.
General: A large species which usually forms large schools but occasionally large adults are observed singly.
Hunts at night and individuals may get separated from others when forming schools for the day.

Sphyraena qenie Klunzinger, 1870

Yellowtail barracuda

Length: To 40 cm.
Distribution: Indo-Pacific, throughout our area.
Depth: 5-50 m.
General: One of the smallest barracudas with a more or less distinct yellow caudal peduncle. Swims in small to large schools along the top of reef slopes in clear coastal and outer reef habitats in relatively shallow depths as adults.

Sphyraena flavicauda Rüppell, 1838

TUNAS & MACKERELS • SCOMBRIDAE

A large commercially important family, comprising several subfamilies, 15 genera and almost 50 species. Fast, streamlined bodies, naked, partly naked or covered with small scales, and head pointed. Series of finlets behind dorsal and anal fins, and 2 or more small keels on each side of caudal peduncle. Swift mostly open water predators which hunt other fishes, cephalopods or crustaceans. Some hunt in small packs or solitair along coastal reefs but the majority of species occur well off shore, sometimes in huge numbers, and associate with current and temperatures. Unlike most fishes which have a body temperature equal to surroundings, some tunas have a body temperature several degrees above. The large schools are commercially targetted for, and form the bases of the world's largest fishing industry with annual catches in excess of 5 million tonnes. All other species are commercially fished for and many of the larger species are favorite gamefish. They occur world-wide and most species range over a vast areas, and some annually migrate extensively.

Oriental bonito

Length: To 1 m.
Distribution: Widespread tropical to subtropical Indo-Pacific, but the cooler zones of our area.
Depth: 0-40 m.
General: Distinctly striped in upper half, fading to mid-line, none in lower half.
A pelagic species, travelling in large schools, often coming close to shore and along reefs to hunt. Feed on small pelagics and break the surface during feeding frenzies.

Sarda orientalis (Temminck and Schlegel, 1844)

Skipjack tuna

Length: To 1 m.
Distribution: Cosmopolitan in tropical to warm-temperate zones. Depth: 1-40 m.
General: Distinctly striped on lower sides, further silvery.
A cosmopolitan species in tropical to warm temperate seas.
Oceanic and coastal, fast swimming in schools, but very wary and almost impossible to photograph underwater.
Commonly reaches fish markets.

Katsuwonus pelamis (Linnaeus, 1758)

TUNAS & MACKERELS • SCOMBRIDAE

Mackerel tuna

Length: To 1 m.
Distribution: Widespread tropical to subtropical Indo-Pacific, including all of our area.
Depth: 0-40 m.
General: Silvery with bluish back, typically a series of diagonal stripes dorsally from below middle of 1st dorsal fin to caudal peduncle.
Pelagic, often making flying visits along coastal reefs with reef fishes diving for cover. Easily confused with the superficially similar Frigate mackerel, *Auxis thazard* an off-shore species which is more slender and has widely separated dorsal fins.

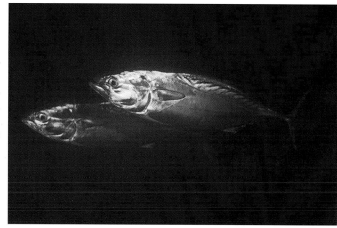

Euthynnus affinis (Cantor, 1850)

Dogtooth tuna

Length: To 1.8 m.
Distribution: Widespread tropical Indo-Pacific, including all of our area.
Depth: 3-100 m.
General: Young often seen singly or a few together over the top of reef flats or in shallow coastal waters. Adults along outer reef drop-offs singly or in small schools, hunting other fishes, often taking reef planktivores which went a little too far out to feed.

Gymnosarda unicolor (Rüppell, 1838)

Bigmouth mackerel

L: To 1 m. Di: Widespread tropical Indo-Pacific. De: 1-40 m.
Ge: A small species in coastal plankton rich waters where large schools swim quickly along reefs until encountering food, then all open their mouth wide and simple swim through, filtering the water with the numerous gill rakers.

Rastrelliger kanagurta (Cuvier, 1817)

289

TUNAS & MACKERELS • SCOMBRIDAE

Double-line mackerel

Length: To 1 m.
Distribution: Widespread tropical Indo-Pacific, including all of our area.
Depth: 2-40 m.
General: A solitairy species, occasionally seen along drop-offs and with seamounts, or near bommies and pinnacles in deep lagoons.
Feeds on small pelagic fish and takes large zooplankton crustacea.

Grammatorcynus bilineatus (Rüppell, 1836)

Spanish mackerel

Length: To 2.2 m.
Distribution: Widespread tropical Indo-Pacific, including all of our area.
Depth: 10-100 m.
General: One of the great hunters, very fast and large, which makes it a popular fish with the game-fisherman.
A common open water species and usually seen solitair, cruising along slopes and drop-offs adjacent to deep water.

Scomberomorus commerson (Lacépède, 1800)

King mackerel

Length: 1.2 m.
Distribution: Northern Australia to New Guinea, Indonesia, Malaysia.
Depth: 5-100 m.
General: Similar to spanish mackerel but deeper bodies and juveniles with broad banding over upper half of body which break up into thin lines and spots with age.

Scomberomores semifasciatus (Macleay, 1884)

TUNAS & MACKERELS • SCOMBRIDAE

Wahoo

Length: To 2.1 m.
Distribution: All tropical to subtropical seas, throughout our area.
Depth: 3-80 m.
General: Oceanic species, occasionally sighted near reefs. Singly or in groups, swimming high in the water colomn. Commercial species reaching about 1.5 m in our area.

Acanthocybium solandri (Cuvier, 1831)

Yellowfin tuna

Length: To 2 m.
Distribution: All open tropical and subtropical seas.
Depth: 1-100 m.
General: An oceanic species which migrate over vast areas in relation to currents and temperatures. They usually are in schools but single individuals are occasionally sighted.
A commercial species and heavily targetted by fisheries worldwide.

Thunnus albacares (Bonnaterre, 1788)

BILLFISHES • ISTIOPHORIDAE

Sailfish

Length: To 3.6 m.
Distribution: Tropical Oceans.
Depth: 3-180 m.
General: A fast oceanic hunter which travels great distances, occasionally near reefs adjacent to deep water.
Other relatives in this small family are the Marlins and spearfish, about 10 species in all, which are wideranging epipelagic species. The marlins are largest, reaching 4.5 m and a weight of over 800 kg.

Istiophorus platypterus (Shaw & Nodder, 1791)

LEFT-EYED FLOUNDERS • BOTHIDAE

A very large family with 15 genera and 90 species in the Indo-Pacific alone. Mostly moderately deep bodied fishes, extremely compressed, with the eyes on left side of the head. Ocular side is pigmented to match surroundings, extremely well camouflaged, used as top-side, and blind side unpigmented is used as underside. Primarily a tropical family, and the number of species rapidly decreases towards temperate zones.

The planktonic larval stage swims upright, initially is bilaterally symmetrical, has an eye on each side. It changes at various stages depending on the genus and reaches different sizes before settling on the substrate, some species in excess of 10 cm. As all flatfishes once settled, they are typically benthic, much of the time buried in the substrate with just eyes exposed. Most species are active on dawn or dusk, a few are nocturnal, and a very few are strictly diurnal. All are carnivorous, feeding on various benthic invertebrates and small fishes.

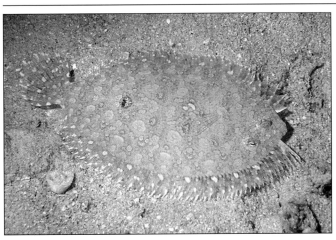

Leopard flounder
L: To 30 cm. Di: Indo-Pacific, throughout our area.
De: 1-100 m. Ge: Males develop long filaments in pectoral fin on ocular side, reaching caudal fin. Shallow coastal reef flats in sand patches ranging to very deep off-shore sand flats.
Below: Head close-up.

Bothus pantherinus (Rüppell, 1830)

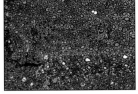

Peacock flounder
L: To 45 cm. Di: Widespread tropical Indo-Pacific. De:1-85 m. Ge: Ocular pec-toral fin large with greatly extended by filamentous rays. Eyes far apart in large adults. Shallow coastal reefs and lagoons, on sand but often in rocky areas, to deep off shore on sand.
Below: Close-up of head.

Bothus mancus (Brousonet, 1782)

SOLES • SOLEIDAE

A large family of small sized fishes, comprising about 30 genera and more than 100 species, however presently poorly defined and in need of revision. Elongate to deep bodied fishes with both eyes on right side. Left side unpigmented and used as underside. Right side is used as top side and has striped or spotted patterns. The head is small, eyes very small, often numerous papillae on blind side. Fins consist entirely of soft rays, almost surrounding body outline, except head ventrally.

Some species are known to possess toxin in small sacs at bases of fin-rays, which can be released to deter predators. Benthic fishes, usually bury in fine sand or mud, feeding on small invertebrates and fishes, sometimes congregating in large numbers.

Eggs are pelagic and larvae symmetrical. Eye migrates over or through head depending on species and postlarvae are to about 25 mm in length when settling.

Peacock sole

Length: To 22 cm.
Distribution: West Pacific and east Indian Ocean to Sri Lanka, including all of our area.
Depth: 3-40 m.
General: Series of toxin glands along dorsal and anal fin bases, their pores visible. When disturbed it releases the toxin which appears milky like from almost the entire out line of the body, and seems to stun predators. Shallow coastal sand or mud flats, usually buried during the day with just outline visible.
The small yellow spots over the body between coarser pattern identifies this species.

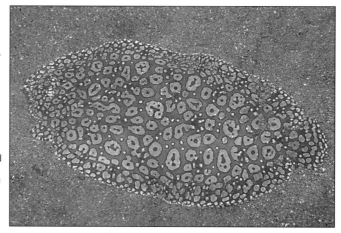

Pardachirus pavoninus (Lacépède, 1802)

Black-spotted sole
L: To 12 cm. Di: Indonesia, only known from Sulawesi and Ambon. De: 1-10 m. Ge: Series of toxin glands along dorsal and anal fin bases, their pores visible. Black spots and blotches to suit habitat. Buried next to rocks, in protected bays. Below White mimic sole, one of several new species.

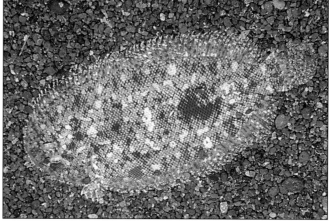

Pardachirus melanospilos (Bleeker, 1854)

TRIGGERFISHES • BALISTIDAE

A moderately large family, comprising 12 genera and about 40 species, occurring world wide, but primarily tropical Indo-Pacific. Large-headed fishes, first dorsal fin with 3 spines, of which 1st stout and lockable in upright position by 2nd spine, 3rd spine small. Pectoral fin usually small, paddle-like. Ventral fin rudimentary, encased with small scales, obvious, protruding from posterior end of ventral flap. Jaws very strong, each with 8 long protruding close-set insiciform teeth. Most species are distinctly coloured and readily identified. They have pelagic larvae, a few are completely pelagic, and some as juvenile only. Benthic juveniles are distinctly coloured and usually differ greatly from adults. In most species there is virtually no differences between sexes. Males of some species are known to make nests and guard eggs vigorously, often attacking divers which venture into their territory. Many species are solitair, found on or near reefs, however nesting sites may consist of numerous males. Only the pelagic-oceanic triggerfish *Canthidermis maculatus* schools. To escape danger, triggerfish use narrow holes or ledges in which they wedge themselves head inwards, with the 'trigger' dorsal spine locked in upright position, making it very difficult to be pulled out. Diet comprises of hard-shelled molluscs and echinoderms.

Clown triggerfish

L: To 35 cm. Di: Indo-Pacific, throughout our area.
De: 5-60 m. Ge: Readily identified by the large round white spots. Adults primarily along deep drop-offs, swimming openly about. Juveniles (below) secretive in caves, and usually deep.

Balistoides conspicillum (Bloch and Schneider, 1801)

Giant triggerfish

Length: To 75 cm.
Distribution: Indo-Pacific, throughout our area.
Depth: 5-35 m.
General: The largest triggerfish and one which can be very aggresive when guarding nests but may attack divers for no obvious reason.
Best is to swim around large individuals at distance, rather than crossing its area. Usually seen solitair on coastal slopes adjacent to deep water **(see opposite page photo from the Andaman Sea).**

Balistoides viridescens (Bloch & Schneider, 1801)

TRIGGERFISHES • BALISTIDAE

Blue triggerfish
L: To 55 cm. Di: Indo-Pacific, throughout our area.
De: 3-50 m. Ge: Deep reef outcrops, frequently over sand to feed on benthic invertebrates, blowing the sand to expose prey.
The name Blue Triggerfish relates to large adult male. Small juvenile shown below.

Pseudobalistes fuscus (Bloch and Schneider, 1801)

Yellow-margin triggerfish
L: To 60 cm. Di: Indo-Pacific, throughout our area.
De: 2-50 m. Ge: Coastal reef slopes, adults solitair, except when nesting on sand patches where they congregate but spread out with each its own territory. Below: Small juvenile.

Pseudobalistes flavomarginatus (Rüppell, 1829)

Starry triggerfish
L: To 55 cm. Di: Indo-Pacific, throughout our area.
De: 10-120 m. Ge: Body pattern readily identifies the species. Adults over sand staying loosely in touch with reefs.
Juvenile (below) in muddy habitats with reef out crops.

Abalistes stellatus (Lacépède, 1798)

TRIGGERFISHES • BALISTIDAE

Half-moon triggerfish

Length: To 30 cm.
Distribution: Indo-Pacific, throughout our area.
Depth: 1-20 m.
General: Adults on shallow reef flats, usually solitair, but several may share a section of reef. Juveniles solitair and shy, quickly diving in holes nearby which are self-made by shifting sand from under rocks.

Sufflamen chrysopterus (Bloch and Schneider, 1801)

Boomerang triggerfish

Length: To 30 cm.
Distribution: Indo-Pacific, throughout our area.
Depth: 3-90 m.
General: Juvenile similar to adult, vertical 'V' pattern from pectoral fin base a diagnostic pattern. Primarily along outer reef slopes and crests, juveniles in surge zones.

Sufflamen bursa (Bloch and Schneider, 1801)

Striped triggerfish
L: To 30 cm. Di: Indo-Pacific, all of our area. De: 6-50 m.
Ge: Mainly at moderate depths in habitats with mixed coral and sponge growth.
Feeds on a great variety of invertebrates, including corals, echinoderms and worms.
Male lacking the stripes over the top of snout (below).

Balistapus undulatus (Mungo Park, 1797)

TRIGGERFISHES • BALISTIDAE

Rhinecanthus cinereus (Bonaterre, 1891)

Strickland's triggerfish

Length: To 25 cm.
Distribution: Indian Ocean from Mauritius to Andaman Sea.
Depth: 25 m.
General: A rare species, known before only from the Mascarenes. For the first time photographed in its natural habitat and documented in the Andaman Sea by Mark Strickland.

Rhinecanthus rectangulus (Bloch and Schneider, 1801)

Wedge-tail triggerfish
L: To 25 cm. Di: Indo-Pacific, in all our area. De: 1-20 m.
Ge: Little change from juvenile to adult, body pattern readily identifies the species. A common shallow reef fish, solitair and territorial, usually just below the intertidal zone. Juveniles in shallow often surge prone reef flats.

Rhinecanthus verrucosus (Linnaeus, 1758)

Black-blotch triggerfish

Length: To 25 cm.
Distribution: Indo-Pacific, throughout our area.
Depth: 0.5-20 m.
General: Shallow coastal reef flats, often adjacent to shoreline mangroves, on mixed sand and hard substrate. Make holes under solid object by shifting sand and quickly retreats when approached.

TRIGGERFISHES • BALISTIDAE

Gilded triggerfish
L: To 22 cm. Di: West Pacific and east Indian Ocean.
De: 20-150 m. Ge: Clear outer reef drop-offs. Forming small to large aggregations in very deep zones. Feed away from wall on zooplankton in the currents. Males differ with the blue cheek and orange fin margins. Below: Female.

Xanthichthys auromarginatus (Bennett, 1831)

Paddlefin triggerfish

L: To 35 cm. Di: Indo-Pacific.
De: 5-60 m. Ge: Clear coastal to outer reef drop-offs. Near ledges with rich invertebrate growth. Post-larvae unusual large, reaching 15 cm at times (see below).

Melichthys vidua (Solander, 1844)

Indian triggerfish

Length: To 25 cm.
Distribution: Indian Ocean, ranging east to Bali, Indonesia.
Depth: 5-35 m.
General: Exposed coastal rocky reefs or outer coral reefs in surge channels with limited coral and algae growth. Stays close to the substrate near entrance of its dug-out hole.
The very similar *M. niger,* circumtropical, occurs in some oceanic locations in our area, it lacks the white margin on tail and stripe on cheek.

Melichthys indicus Randall & Klausewitz, 1973

FILEFISHES • MONACANTHIDAE

A large family with small to medium sized fishes, comprising about 30 genera and an estimated 100 species, found in all major oceans. Particularly well represented in Australia where they are known as leatherjackets, with 27 genera and almost 60 species. Japan and South Africa have about 16 species each.

Most species change in shape with age, juveniles may by almost circular and very compressed, whilst the large adults are very elongate. The body is covered by tiny prickly scales, forming a tough leathery or velvet-like skin, and only in some species the scales are visible by the naked eye. A prominant and separate 1st dorsal fin spine is an obvious feature in most species, often very long, or armed with series of downward directed barbs on edges, usually followed by a 2nd much smaller embedded spine and narrow membrane. In nearly all species the spine is lockable in upright position, and often when folded back fitting into a shallow to deep groove.

Little is known about the spawning for most species. Some of the subtropical filefishes release eggs in open water, however in some species they sink, thus eggs are scattered over the substrates (usually seagrass beds) and in others they float, thus pelagic. A few of the smaller coral dwellers have nests and guard eggs like the closely related triggerfishes. Larvae are pelagic and postlarvae either settle when about 10-15 mm in most species, some juveniles have a long pelagic stage and a few adults are pelagic as well.

Most of the reef species are distinctly coloured or shaped and identification usually easy, however a great number of species associate with weeds or seagrasses in which they camouflage. Most are greenish and some species are very difficult to identify to a species underwater or from a photograph, especially when juvenile.

Weedy filefish

Length: To 18 cm.
Distribution: Indo-Pacific, including all of our area.
Depth: 1-25 m.
General: The weedy look from the skin flaps varies in density. Secretive in weed areas and extremely well camouflaged and often found by accident when disturbing the weed it is in.

Below: a juvenile biting onto soft coral as anchorage at night. Biting onto things at night, usually weeds, is typical behaviour for many members in the family.

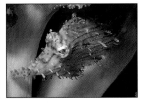

Chaetoderma penicilligera (Cuvier, 1817)

FILEFISHES • MONACANTHIDAE

Long-nose filefish
L: To 10 cm. Di: Indo-Pacific, including all of our area.
De: 0.5-35 m. Ge: Coastal to outer reef lagoons, anywhere where large patches of *Acropora* corals are present. Adults pair whilst juveniles swim in small groups between or over the tips of the corals, feeding primarily on the polyps.

Oxymonacanthus longirostris (Bloch & Schneider, 1801)

Mimic filefish

Length: To 10 cm.
Distribution: Indo-Pacific, including all of our area.
Depth: 1-25 m.
General: Shallow protected reefs to about 25 m, openly about, often in small aggregations. Mimics *Canthigaster valentini* a poisonous puffer in colour and shape in great detail, and shares the same habitat. It differs in having a 1st dorsal fin, and long-based 2nd dorsal and anal fins.

Paraluteres prionurus (Bleeker, 1851)

Briste-tail filefish

Length: To 70 mm.
Distribution: Indo-Pacific, including Thailand, Malaysia, Vietnam, Philippines and Indonesia.
Depth: 1-25 m.
General: Coastal sand flats or slopes with small patches of weeds, large algaes on rock, or with growth on jetty pilons.

Acreichthys tomentosum (Linnaeus, 1758)

FILEFISHES • MONACANTHIDAE

Pervagor janthinosoma (Bleeker, 1854)

Ear-spot filefish

Length: To 13 cm.
Distribution: Indo-Pacific, including all of our area.
Depth: 0.5-15 m.
General: Variable from pale green to dark brown, caudal fin usually orange and distinct vertical black blotch above pectoral fin base.
Shallow reef flats and in rocky estuaries, very secretive amongst boulders, sponges and often on jetty pilons. Singly or in pairs.

Pervagor melanocephalus (Bleeker, 1854)

Black-head filefish

Length: To 10 cm.
Distribution: Philippines, Malaysia, Indonesia to Samoa and Great Barrier Reef.
Depth: 10-45 m.
General: Clear coastal to outer reefs, usually fairly deep in rich invertebrate habitats with mixed soft corals and sponges, secretive in the shade of the reef. Usually in pairs, and easily identified by the orange body and dark head.

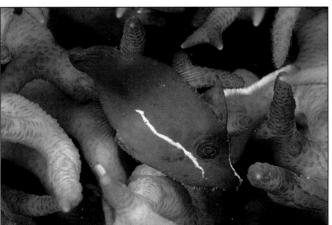

Pervagor nigrolineatus (Herre, 1927)

White-line filefish

Length: To 80 mm.
Distribution: West Pacific, ranging to east Indian Ocean, probably in all of our area.
Depth: 0.5-25 m.
General: Very shallow protected reef flats in dense coral heads such as bushes of *Seriatopora* species. Small groups moving through inside where they pick on the coral for food, either picking off small crustacea and/or feeding on poplys. Best recognised by the long white line on their side and over the top of snout.

FILEFISHES • MONACANTHIDAE

Peron's filefish

Length: To 25 cm.
Distribution: Indonesia, Bali to northern Sulawesi, and northern Australia.
Depth: 1-25 m.
General: Clear coastal and protected inner reefs with mixed rock, coral and algae habitat. Very similar to next species but has larger tail and male lacks the spine ridge on the sides of caudal peduncle which is prominent in *P. macrurus*.

Pseudomonacanthus peroni (Hollard, 1854)

Strapweed filefish

Length: To 45 cm.
Distribution: Throughout Indonesia, to Papua New Guinea and tropical Australia.
Depth: 1-15 m.
General: Found primarily in broad-leaf seagrass beds, coastal as well as small remote islands with protected bays with suitable habitat.
A common species but hides and is well camouflages. Often swims in pairs, male has spiny ridge on the sides of the caudal peduncle.

Pseudomonacanthus macrurus (Bleeker, 1856)

Short-snout filefish

L: To 11 cm. Di: Uncertain, Indo-Pacific. De: 6-50 m.
Ge: Several similar species, usually referred to as *P. japonicus*, and this species usually called *P. cryptodon*. Males have convex snout.
Below: Juvenile.

Paramonacanthus curtorhynchus (Bleeker, 1855)

FILEFISHES • MONACANTHIDAE

Rhinoceros filefish

L: To 18 cm. Di: Indo-Pacific, throughout our area.
De: 3-50 m. Ge: Shallow coastal reefs and estuaries with seagrasses and soft coral in tropical waters. Secretive, often positioned vertically with head down, behind weeds or seapens on open substrates.

Pseudaluterus nasicornis (Temminck and Schlegel, 1850)

Scribbled filefish

Length: To 1 m (including very large tail).
Distribution: Circumtropical, throughout our area.
Depth: surface to 80 m.
General: Small juveniles yellow with black spots, becoming ornamented with bright blue with age. Juveniles pelagic, up to a large size, with floating weeds and forming schools, adults usually solitair and only occasionally seen in small groups on coastal reefs.

Aluterus scriptus (Linnaeus, 1758)

Unicorn filefish

Length: To 75 cm.
Distribution: Circumtropical, throughout our area.
Depth: surface to 50 m.
General: A totally pelagic species which as small juveniles drift with floating weeds far off shore and swim with some of the faster moving jellies which sometimes brings them close to reefs. Adults form schools and may be seen under large sargassum rafts.

Aluterus monoceros (Linnaeus, 1758)

BOXFISHES • OSTRACIIDAE

This family comprises some unusual species, known as box or trunkfishes and cowfishes, in 6 genera and 20 species, distributed in all major tropical oceans. The body is largely covered with well defined hexagonal bony plates, fused into a ridged case with holes for the moving parts, the fins, caudal peduncle, mouth, eyes and gills. The genera differ primarily in cross-section shapes, with flat to rounded bellies, tapering above slightly to almost a square or a broad triangle with a dorsal ridge. The surfaces are smooth but some species have prominant spines above eyes or on back.

The body is covered with a toxic mucus, which deter predators but also the toxin is released when under stress, killing other fish or themselves if kept confined. Some species are sexually dimorphic with colourful males, the latter derives from females. To spawn, pairs encircle each other and swim upwards high above the substrate, sometimes to surface, to release their pelagic eggs. Postlarvae are only about 10 mm and almost square, often referred to as dice fish.

Longhorn cowfish

Length: To 50 cm.
Distribution: Indo-Pacific, throughout our area.
Depth: 1-50+ m.
General: Body short in juveniles, cube-like when just settled, and elongate as large adults. Horn-like spines develop quickly after settling, in adults very large.
Caudal fin increasing in length with age, reaching almost body length. Mostly in sandy lagoons, feeding on sand-invertebrates and exposing them by blowing sand away.

Lactoria cornuta (Linnaeus, 1758)

Thorny-back cowfish

Length: To 20 cm.
Distribution: Indo-Pacific, throughout our area.
Depth: 3-50 m.
General: Green to brown with blue spots and scribbles. Coastal rocky reef and estuaries, usually solitair on sand slopes with mixed sponge and weed habitats.
Best identified by prominent spine in middle of back.

Lactoria fornasini (Bianconi, 1846)

BOXFISHES • OSTRACIIDAE

Round-belly cowfish

Length: To 30 cm.
Distribution: Indo-Pacific, throughout our area.
Depth: 1-50+ m.
General: Dorsal profile straight from eyes to caudal fin, ventral profile deeply rounded, particularly in juveniles, which identifies this species. Rocky, algae rich reefs to about 30 m depth. Juveniles may have prolonged pelagic stage in some areas.

Lactoria diaphana (Bloch and Schneider, 1801)

Solor boxfish
L: To 12 cm. Di: Indonesia, Philippines and further east.
De: 3-40 m. Ge: Sheltered rich coral reef crests, slopes and in caves along deep outer reef drop-offs. Juveniles with black stripes which break-up with growth and form a maze pattern on top and sides. Female below.

Ostracion solorensis Bleeker, 1853

Black boxfish
L: To 20 cm. Di: Indo-Pacific, throughout our area.
De: 3-35 m. Ge: Juvenile and female black with small white spots. Males derive from females. Female below.

Ostracion meleagris Shaw, 1796

306

BOXFISHES • OSTRACIIDAE

Cube boxfish

L: To 45 cm. Di: Indo-Pacific, throughout our area.
De: 1-45 m. Ge: Small juveniles cube-shaped, elongating with growth to moderately elongate. Caudal fin increasing in size with age and caudal peduncle becomes long and thick.

Ostracion cubicus Linnaeus, 1758

Shortnose boxfish

Length: To 30 cm.
Distribution: Indo-Pacific, throughout our area.
Depth: 10-50 m.
General: A rare species, usually mistaken for another species. It has a unique colour pattern and large adult develop a small but prominent hump on the snout which is either smaller in the species above or very large in the species below.

Ostracion nasus Bloch, 1785

Rhino boxfish

Length: To 50 cm.
Distribution: Indo-Pacific, throughout our area.
Depth: 10-50+ m.
General: A deep water species, only occasionally encountered in divable depths, and only in areas adjacent to such habitat. Best identified by the large hump on the snout, but there is similar species in the Indian Ocean has ranging to Indonesian waters which has a small hump on the tip of the snout.

Ostracion rhinorhynchus Bleeker, 1852

PUFFERFISHES • TETRAODONTIDAE

A large family comprising about 20 genera and at least 100 species, consisting of several distinct groups and sub-families. The sharp-nosed puffers, Canthigasterinae, and the short-nosed species, Tetraodontinae, the latter however comprises the large species as well as a great variety of small coastal species. Their common name applies to their ability to inflate themselves to almost balloon proportions to deter predators. In addition many species are prickly which when inflated become even more hazardous to consume, and all species are poisonous in either internal organs or skin toxins, thus in all the message should be clear: leave me alone. Some species are clearly marked to advertise their likely fatal properties and some totally unrelated species have taken advantage of this by copying the image in both, shape and colouration, thus mimicry, to bluff potential predators with their looks.

They produce tiny round eggs, less than 1 mm in diameter, which are sticky and demersal. Larvae are pelagic, up to about 6 mm long when getting ready to settle on the substrate. With some species diet comprises almost everything whilst others have preferences for certain invertebrates or algaes, but all seem to consume a great variety.

Saddled puffer

Length: To 10 cm.
Distribution: Indo-Pacific, throughout our area.
Depth: 1-20 m.
General: A distinct species and serves as model for non-poisonous mimics, a mona-canthid *(Paraluterus prionurus)* and a juvenile serranid *(Plectropomus leavis)*. Common on shallow coastal to outer reefs, often in pairs, feeding on algae and a variety of small invertebrates. Males have blue lines radiating from the eye.

Canthigaster valentini (Bleeker, 1853)

Crowned puffer

Length: To 14 cm.
Distribution: Indo-Pacific, throughout our area but two forms.
Depth: 10-80 m.
General: Similar to the saddled puffer but more colourful and usually found on deep rubble slope. Usually in pairs, slowly moving near the substrate, feeding near or on sponges on a variety of invertebrates. Indian Ocean form may represent a different species.

Canthigaster coronata (Vaillant & Sauvage, 1875)

PUFFERFISHES • TETRAODONTIDAE

Circle-barred puffer

Length: To 75 mm.
Distribution: Philippines, Indonesia, and ranging to eastern Australia and Fiji.
Depth: 15-40 m.
General: A secretive species, usually seen solitair in caves along deep drop-offs on outer reefs, but several were seen together in a loose group on a deep coastal reef flat with numerous plate corals. Although fairly common, rarely observed by divers except by those using a light during the day to inspect caves.

Canthigaster ocellicincta Allen & Randall, 1977

Grey-top puffer

Length: To 10 cm.
Distribution: Widespread tropical west Pacific, ranging west to Bali.
Depth: 10-40 m.
General: A rare species in our area, deep in excess of 20 m and solitair. Rich rubble slopes with sponges and soft corals or along deep drop-offs on the bottom of large open caves. Some variation in colour, often dark on top from the snout to tail.

Canthigaster epilampra (Jenkins, 1903)

Fine-spotted puffer

Length: To 12 cm.
Distribution: West Pacific to Indian Ocean, throughout our area.
Depth: 1-30 m.
General: A coastal species, usually in pairs on sand or mud and in seagrass beds along the edges to sand. The smaller spots compared to other similar species at similar size identify this species. The spots are proportionally much larger and less numerous in juveniles.

Canthigaster compressa (Procé, 1822)

PUFFERFISHES • TETRAODONTIDAE

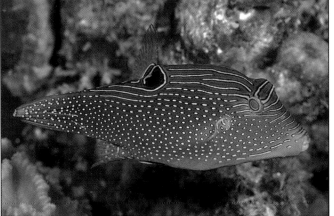

False-eye puffer
L: To 10 cm. Di: Indonesia to Palau tropical Indo-Pacific. De: 6-50 m. Ge: Along drop-offs in open caves but also on clear water coastal reef crests. Juveniles similar but more lined and with growth the lines become more numerous and break up into spots. Replaced by *C. solandri* (below) in some areas.

Canthigaster papua Bleeker, 1848

Ambon puffer

Length: To 12 cm.
Distribution: Indo-Pacific, throughout our area.
Depth: 1-30 m.
General: Combination of spots and lines identifies this species. A solitary species, secretive in reefs as juveniles and rarely seen. Adults in high energy surge zones, usually seen dashing from one shelter to the next.

Canthigaster amboinensis (Bleeker, 1865)

Bennett's puffer
L: To 9 cm. Di: Indo-Pacific, throughout our area.
De: 1-30 m. Ge: Colour variable with habitat, greenish in weedy areas and brownish above on rocky or coral reefs. Usually in pairs as adult and mainly in coastal bays.

Right: The black-spotted puffer rests during the day in a leather-coral off Bali, Indonesia.
See also page 313.

Canthigaster bennetti (Bleeker, 1854)

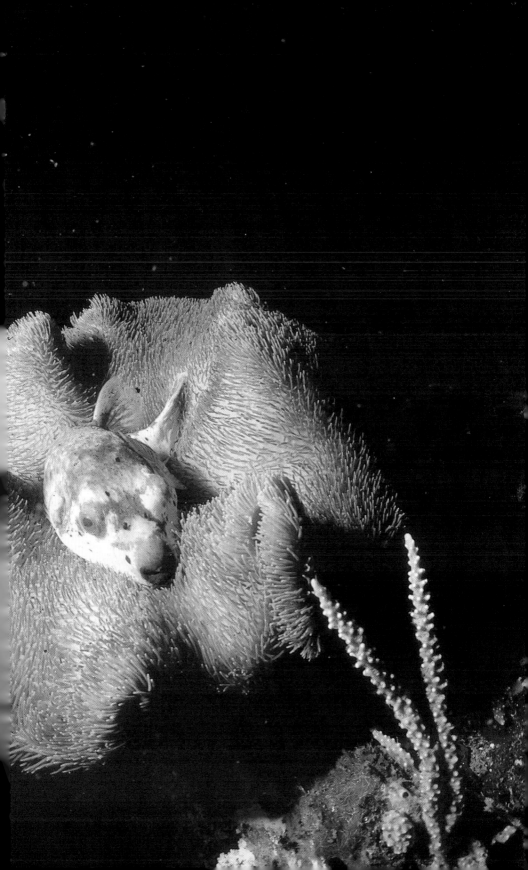

PUFFERFISHES • TETRAODONTIDAE

Arothron manilensis (Procé, 1822)

Manila puffer

Length: To 30 cm.
Distribution: West Pacific and eastern Indian Ocean, containing all of our area.
Depth: 1-25 m.
General: Variable dark grey to yellowish, with a few to many longitudinal lines. Coastal mud and sand flats and slopes, and in seagrass areas. Juveniles enter brackish water and lower reaches of fresh water streams. Adults singly or in pairs and my mix with the next species in some areas.

Arothron immaculatus (Bloch & Schneider, 1801)

Yellow-eye puffer

Length: To 30 cm.
Distribution: Indo-Pacific, throughout our area.
Depth: 1-25 m.
General: A coastal species found mainly in muddy areas and in the vicinity of seagrass beds. Sleeps most of the days in shallow depressions in the bottom or against logs. In Flores commonly mixes with above species, sometimes in small aggregations.

Arothron stellatus (Bloch and Schneider, 1801)

Starry puffer

Length: To 1.2 m, usually less than 1 m.
Distribution: Indo-Pacific, including all of our area.
Depth: 5-60 m.
General: Coastal reefs, juveniles often with land debris on the substrate such as palm leaves or logs.
Adults on patch reefs with caves and large sponges, usually seen solitair. Feeds on invertebrates, mainly echinoderms.

PUFFERFISHES • TETRAODONTIDAE

Black-spotted puffer

Length: To 30 cm.
Distribution: Indo-Pacific, including all of our area.
Depth: 3-25 m.
General: The only species in this genus found mainly on coral reefs, often sleeping in barrelsponges.
Highly variable, grey to brown or bright yellow to orange, black-spotted to various degrees, and sometimes a mixture of colours. Occurs singly or in pairs. See also page 311.

Arothron nigropunctatus (Bloch & Schneider, 1801)

Reticulated puffer

Length: To 45 cm.
Distribution: Tropical Indian Ocean to Indonesia.
Depth: 1-25 m.
General: A coastal species found mainly in estuaries and protected muddy bays. Juveniles in mangroves and entering lower reaches of streams. Best separated from similar species by the striations below the head and belly.

Arothron reticularis (Bloch & Schneider, 1801)

PUFFERFISHES • TETRAODONTIDAE

Arothron mappa (Lesson, 1830)

Mappa puffer

Length: To 60 cm.
Distribution: Widespread tropical Indo-Pacific, including all of our area.
Depth: 5-30 m.
General: Coastal to outer reefs. A solitary species in caves and under table corals, and although common in some areas secretive thus not often observed. In Java it is the most common large puffer. Highly variable species with scribbles, wire-netting patterns or lines, but usually numerous lines radiating from the eye.

Arothron hispidus (Linnaeus, 1758)

White spotted puffer

Length: To 50 cm.
Distribution: Widespread tropical Indo-Pacific, including all of our area.
Depth: 3-50 m.
General: A variety of habitats from coastal estuaries to outer reef slopes, but mainly in sandy areas between reefs or sea-grass beds. Best recognised by the white circles around pectoral base and eye.
Feeds on virtually every marine organism, dead or alive.

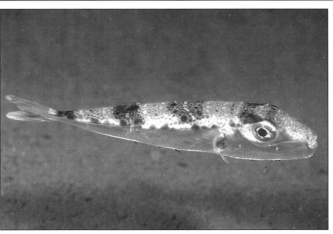

Lagocephalus sceleratus (Gmelin, 1788)

Silver puffer

Length: To 80 cm.
Distribution: Indo-Pacific, including all of our area.
Depth: 0-25 m.
General: Juveniles in coastal bays, adults coastal to pelagic forming schools.
In some areas dangerous to divers and swimmers, and is most feared in western Australia. It bites off toes, through human bone and even fish hooks. In the addition it has poisonous flesh, which made Captain Cook sick.

PORCUPINEFISHES • DIODONTIDAE

A small family of medium to moderate-sized fishes, comprising 6 or 7 genera and about 20 species, occurring primarily in tropical waters of the major oceans, but some range to temperate zones. They are similar to pufferfishes in their ability to inflate themselves to considerable size, but have in addition large spines which point outwards when inflated and teeth are totally fused into beaklike jaws. The spines are modified scales and maybe rigid or movable, or both.

All species have pelagic eggs and juveniles and one species is fully pelagic. Adults are benthic and occur on shallow sheltered reefs or deeper off-shore ranging to depths of more than 300 m. Most species appear to be nocturnal, usually sheltering in caves during the day, and feed on a variety of invertebrates, including hard-shelled species. Although reputed to be poisonous, some species are eaten by islanders in the Pacific with no obvious ill-effects.

Fine-spotted porcupinefish

Length: To 30 cm.
Distribution: Circumtropical, ranging to subtropical zones and throughout our area.
Depth: 3-100 m.
General: Spines long and narrow, erectile, many spines longer than eye-diameter, and longest on forehead.
Sheltered bays and estuaries, but also known from trawls.

Diodon holocanthus Linnaeus, 1758

Masked porcupinefish

Length: To 45 cm.
Distribution: Indo-Pacific, throughout our area.
Depth: 1-90 m.
General: Spines of moderate size and narrow, erectile, longest spines above pectoral fins. Juveniles in shallow lagoons between shore and reef barriers, adults usually fairly deep on coastal reefs, in ledges during the day.

Diodon liturosus Shaw, 1804

PORCUPINEFISHES • DIODONTIDAE

Diodon hystrix Linnaeus, 1758

Black-spotted porcupinefish

Length: To 71 cm.
Distribution: Circumtropical, ranging to subtropical zones and throughout our area.
Depth: 5-100+ m.
General: Spines erectile but rather short. Large individuals usually seen solitair, often swimming high above the substrate during the day, but usually found near deep water. Juveniles with floating weeds where they may reach a moderate size before settling on the substrate.

Chilomycterus reticularis (Linnaeus, 1758)

Few-spine porcupinefish

Length: To 55 cm.
Distribution: Circumtropical, ranging to subtropical zones and throughout Southeast Asia.
Depth: 0-100+ m.
General: Spines short and few. Juveniles stay pelagic to a large size, up to 20 cm, with floating weeds or objects to give cover. Adults move to deep water and are only seen on very deep slopes; and usually at night when they seem to feed in shallower depths.

Cyclichthys orbicularis (Bloch, 1785)

Rounded porcupinefish

L: To 20 cm. Di: Indo-Pacific.
De: 10-50+ m. Ge: A common species, hiding deep in reefs during the day, but active at night One of the smallest species with very large eyes. Mainly on rich coastal reef slopes, sleeping in spouges, below.

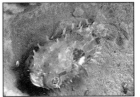

ENGLISH INDEX

Ambon puffer	310
Andaman foxface	271
Andaman sweetlips	154
Arrow-head soapfish	110
Badger clingfish	52
Bald-spot spinecheek	159
Banda cardinalfish	122
Banded goatfish	163
Banded pipefish	74
Banded snake eel	40
Banded splendour wrasse	234
Banded thicklip	221
Banded trevally	134
Bar-cheek trevally	135
Bar-tail goatfish	164
Bar-tail moray	38
Barramundi cod	93
Barred garfish	55
Barred moray	36
Bartels' dragonet	250
Bath's combtooth blenny	247
Beaked coralfish	184
Bearded rockling	48
Bennett's puffer	310
Big mudgoby	260
Big-eye emperor	157
Big-eye trevally	133
Big-spot blenny	248
Bigmouth mackerel	289
Bignose unicornfish	277
Bird-nose wrasse	226
Black & Yellow damsel	205
Black anemonefish	199
Black boxfish	306
Black garden eel	41
Black snapper	138
Black-back butterflyfish	176
Black-barred butterflyfish	174
Black-base bullseye	166
Black-blotch foxface	271
Black-blotch triggerfish	298
Black-head filefish	302
Black-margin bullseye	167
Black-orange dragonet	251
Black-ray shrimpgoby	256
Black-spot angelfish	196
Black-spot worm goby	268
Black-spotted porcupinefish	316
Black-spotted puffer	311, 313
Black-spotted sole	293
Black-stripe worm goby	268
Black-striped cardinalfish	124
Black-tail humbug	204
Black-tail parrotfish	238
Black-tail sergeant	207
Black-tipped fusilier	148
Blackbelt hogfish	219
Blacktail snapper	141
Blacktip reef shark	18
Bleeker's damsel	208
Bleeker's grouper	98
Blood-drop squirrelfish	61
Blotched bigeye	118
Blotched fantail ray	30
Blotched grinner	46
Blotched hawkfish	212
Blotchfin scorpionfish	82
Blotchy swell shark	16
Bludger trevally	135
Blue & gold angelfish	192
Blue blanquillo	127
Blue lined barracuda	287
Blue razorfish	228
Blue shark	21
Blue shrimpgoby	256
Blue triggerfish	296
Blue-backed angelfish	196
Blue-barred parrotfish	235
Blue-barred ribbon goby	267
Blue-dash butterflyfish	179
Blue-dash fusilier	147
Blue-face angelfish	189
Blue-fin trevally	133
Blue-green puller	205
Blue-head fairy-wrasse	231
Blue-head tilefish	129
Blue-ribbon wrasse	223
Blue-ringed angelfish	188
Blue-speckled rubble goby	260
Blue-spine unicornfish	274
Blue-spotted fantail ray	29
Blue-spotted jawfish	216
Blue-spotted large-eye bream	157
Blue-spotted mask-ray	29
Blue-spotted sand dragonet	252
Blue-spotted sand-diver	239
Blue-stripe spinecheek	159
Blue-striped angelfish	190
Blue-striped snapper	143
Bluelined rockcod	91
Bluespotted rockcod	91
Bluntnose sixgilled shark	10
Boers batfish	172
Boomerang triggerfish	297
Brassy drummer	168
Bridled anemonefish	199
Briste-tail filefish	301
Broad-barred sleepergoby	253
Bronze soldierfish	63
Brooch lanternfish	47
Brown bamboo shark	12
Brown sweetlips	154
Brown tang	285
Brown-spotted grouper	98
Brown-stripe snapper	142
Bull shark	20
Butterfly whiptail	158
Byno goby	255
Chain-lined wrasse	222
Chameleon tilefish	129
Checkered snapper	144
Cheek-bar cardinalfish	123
Cheek-spots cardinalfish	123
Chevron barracuda	286
Chevroned butterflyfish	179
Cigar wrasse	221
Circle-barred puffer	309
Citron butterflyfish	176
Clark's anemonefish	202
Cleaner wrasse	230
Clouded moray	36
Clown anglerfish	51
Clown triggerfish	294
Coachwhip trevally	135
Cobia	130
Cockatoo leaf fish	84
Cockerel wrasse	233
Comet	117
Common blacktip shark	18
Common lionfish	79
Common seahorse	71
Convict blenny	242
Convict surgeonfish	283
Coral beauty	193
Coral cat shark	16
Coral hawkfish	213
Coral rabbitfish	272
Coral rockcod	95
Coral shrimpfish	67
Coronation lyretail-cod	92
Cowtail stingray	26
Crab-eyed goby	253
Crested sabretooth blenny	246
Crimson soldierfish	62
Crocodile fish	86
Crocodile longtom	54
Cross-hatch butterflyfish	174
Crown squirrelfish	59
Crowned puffer	308
Cube boxfish	307
Damsel dwarf-angelfish	192
Dash and dot goatfish	162
Deep-bodied silverbelly	149
Diamond-scale mullet	217
Diana's hogfish	219
Dick's damsel	210
Dogtooth tuna	289
Dotted butterflyfish	183
Double-bar goatfish	163
Double-barred rabbitfish	272
Double-ended pipehorse	72
Double-line mackerel	290
Dragon moray	36
Dusky dottyback	114
Dusky grouper	96
Dusky parrotfish	237
Dusky shark	19
Dusky wrasse	222
Dwarf lionfish	78
Dwarf pipehorse	72
Ear-spot filefish	302
Eastern clown-anemonefish	203
Eastern skunk anemonefish	201
Eclipse hogfish	220
Ehrenberg's snapper	142
Eibl's angelfish	194
Eight-line cardinalfish	121
Ellow-back basslet	100
Ember parrotfish	235
Emperor angelfish	188
Epaulette shark	13
Exquisite fairy-wrasse	230
Eye-line surgeonfish	281
Eye-patch butterflyfish	177
Eye-spot surgeonfish	281
Eyelash harptail blenny	245
False leopard wrasse	221
False three-spot cardinalfish	123
False-eye grubfish	240
False-eye puffer	310
Few-spine porcupinefish	316
Filamented flasher wrasse	232
Filamented sand goby	261
Fine-lined bristletooth	284
Fine-spotted porcupinefish	315
Fine-spotted puffer	309
Finelined squirrelfish	59
Fingered dragonet	252
Firetail devil	115
Five-line cardinalfish	121
Five-line snapper	143
Flag-tail shrimpgoby	258
Flagtail blanquillo	127
Flagtail rockcod	90
Flame dwarf-angelfish	194
Flores basslet	101
Flying gurnard	85
Footballer coralcod	93
Forsten's parrotfish	236
Forster's hawkfish	214
Foursaddle grouper	97
Fringe-lip flathead	86
Gaimard's wrasse	225
Galapagos shark	17
Giant anglerfish	51
Giant fusilier	146
Giant moray	37
Giant squirrelfish	58
Giant sweetlips	154
Giant trevally	133
Gilded triggerfish	294
Glasseye	118
Gold-band fusilier	147
Gold-saddle rabbitfish	270
Gold-specs jawfish	216
Gold-spot emperor	155
Gold-spotted rabbitfish	269
Gold-spotted sweetlips	153
Golden hawkfish	212
Golden moray	37
Golden sergeant	205
Golden spinecheek	160
Golden tilefish	128
Golden trevally	132
Goldring bristletooth	284
Goldspot herring	42
Great barracuda	286
Great hammerhead	23
Green jobfish	137
Green-band sleepergoby	254
Greenback bullseye	166
Greenish damsel	208
Grey bamboo shark	12
Grey nurse shark	15
Grey reef shark	22
Grey-streak lizardfish	44
Grey-top puffer	309
Hairtail blenny	244
Half & half goatfish	164
Half-barred snapper	144
Half-circled angelfish	188
Half-moon triggerfish	297
Harlequin rockcod	91
Harlequin sweetlips	151
Head-band butterflyfish	174
Hector's goby	256
Herald's dwarf-angelfish	191
Highcrest triplefin	243
Highfin amberjack	132
Honeycomb moray	35
Hooded epaulette shark	13
Horned bannerfish	186
Horse-shoe surgeonfish	282
Humpback snapper	139
Humphead parrotfish	238
Humpnose unicornfish	275
Indian flame basslet	106
Indian humbug	204
Indian mimic surgeonfish	279
Indian triggerfish	299
Indian white wrasse	223
Jans's pipefish	75
Jansen's wrasse	227
Japanese angel shark	23
Japanese bullhead shark	11
Java cownose ray	31
Java rabbitfish	270
Jenkins whipray	27
Jewel basslet	104
Jewel damsel	210
Jeweled blenny	249
Keyhole angelfish	193
King mackerel	290
Lacy scorpionfish	82
Lagoon shrimpgoby	259
Lamarck's angelfish	196
Lance blenny	244
Lattice butterflyfish	182
Lemon peel	191
Leopard blenny	249
Leopard flounder	292
Leopard rockcod	90
Leopard whipray	28
Lined soapfish	108
Lined surgeonfish	283
Lined sweetlips	152
Little dragonfish	64
Little priest	43
Little unicornfish	276
Loki whip-goby	264
Long grinner	47
Long-barbel goatfish	163
Long-finned cod	95
Long-finned dottyback	114
Long-nose butterflyfish	184
Long-nose filefish	301
Long-rayed sand-diver	239
Long-snout clingfish	52
Long-snout pipefish	73
Longfin perchlet	100
Longhorn cowfish	305
Longnose hawkfish	211, 214
Longnose unicornfish	276
Lori's basslet	107
Lunartail snapper	141
Luzon basslet	102
Lyre-tail dart goby	265
Lyre-tail grubfish	241
Lyre-tail hawkfish	213
Lyre-tail dottyback	111, 114
Lyre-tail hogfish	218
Mackerel tuna	289
Magenta dottyback	113
Magpie sweetlips	151
Majestic angelfish	189
Malabar snapper	140
Mandarinfish	251
Mangrove jack	140
Mangrove whiptail	28
Manila puffer	312
Manta ray	33
Many-banded angelfish	195
Many-host goby	262
Mappa puffer	314
Marbled dragonet	250
Marbled rockcod	95
Marbled torpedo ray	26
Masked porcupinefish	315
Masked rabbitfish	272
Maze rabbitfish	270
Metalic shrimpgoby	258
Midnight snapper	138
Mimic filefish	301
Mirror basslet	103
Moluccan cardinalfish	125
Moluccen snapper	145

Name	Page	Name	Page	Name	Page	Name	Page
Moonbeam dwarf-angelfish	192	Red-band basslet	103	Slender suckerfish	130	Variegated jawfish	216
Moorish idol	273	Red-barred rockcod	98	Slender unicornfish	275	Variegated lizardfish	45
Mouth-fin squirrelfish	61	Red-head fairy-wrasse	231	Slingjaw wrasse	234	Velvet surgeonfish	280
Moyer's dragonet	251	Red-marbled lizardfish	44	Small-spot goatfish	162	Vermicular coralcod	94
Mud-reef goby	261	Red-margin shrimpgoby	258	Small-spotted grouper	96	Vermiculate angelfish	190
Multi-striped cardinalfish	125	Red-ribbon wrasse	227	Small-tooth emperor	156	Very-long-nose butterflyfish	184
Multibar pipefish	74	Red-spots razorfish	229	Smiling shrimpgoby	257	Violet soldierfish	63
Mushroom-coral pipefish	75	Red-spotted blenny	248	Smith's harptail blenny	245	Violet squirrelfish	60
Narrow sawfish	24	Red-stripe basslet	102	Smooth flutemouth	66	Wahoo	291
Narrow-banded wrasse	224	Red-stripe tilefish	128	Smooth hammerhead	23	Warty-lip mullet	217
Nate wobbegong	12	Redcoat squirrelfish	58	Snout-spots grouper	97	Wedge-tail triggerfish	298
Nose-spot cardinalfish	120	Reef stonefish	83	Snowflake soapfish	108	Weedy filefish	300
Oblique-banded sweetlips	152	Regal angelfish	195	Snub-nose dart	136	Western clown-anemonefish	203
Oceanic whitetip shark	21	Reticulated butterflyfish	178	Snubnose drummer	168	Western skunk anemonefish	201
One-fin flashlightfish	57	Reticulated puffer	313	Soft-coral goby	264	Whale shark	9, 10
One-spot snapper	141	Reticulated whipray	28	Soldier lionfish	79	Whiskered pipefish	73
One-stripe featherstar clingfish	53	Rhino boxfish	307	Solor boxfish	306	White spotted puffer	314
Onespine unicornfish	274	Rhinoceros filefish	304	Spanish flag snapper	143	White-banded possum-wrasse	233
Orange rockcod	89	Ribbon eel	34	Spanish mackerel	290	White-blotch razorfish	229
Orange-banded coralfish	183	Rifle cardinalfish	124	Speckled wrasse	224	White-dashed wrasse	224
Orange-blotch parrotfish	236	Right: Pixie wrasse	225	Sphynx goby	255	White-edged lyretail-cod	92
Orange-blotch surgeonfish	278	Rigid shrimpfish	68	Spine-cheek anemonefish	198	White-eyed moray	39
Orange-fin anemonefish	200	Ring-eye pygmy goby	262	Spiny leaf fish	84	White-line filefish	302
Orange-line worm goby	268	Ring-eyed hawkfish	214	Spiny-tail puller	206	White-lined combtooth blenny	247
Orange-lined cardinalfish	124	Ring-tail cardinalfish	122	Splendid dottyback	115	White-lined lionfish	78
Orange-lined sweetlips	153	Ringtail surgeonfish	282	Splendid soldierfish	63	White-margin unicornfish	275
Orange-spine unicornfish	277	Rippled rockskipper	248	Spot-banded butterflyfish	177	White-mouth moray	38
Orange-spotted sleeper-goby	254	Robust fusilier	146	Spot-face moray	39	White-speckled grouper	97
Oriental bonito	288	Robust ghostpipefish	68	Spot-face shrimpgoby	259	White-spot cardinalfish	122
Oriental sweetlips	152	Rockmover wrasse	228	Spot-face surgeonfish	282	White-spotted bamboo shark	13
Ornate butterflyfish	178	Round batfish	170	Spot-tail shark	18	White-spotted eagle ray	32
Ornate ghostpipefish	70	Round-belly cowfish	306	Spotfin lionfish	76	White-spotted guitarfish	25
Oval-spot butterflyfish	180	Round-spot goatfish	164	Spotted coralcod	93	White-square rockcod	92
Pacific double-saddle butterflyf.	181	Rounded porcupinefish	316	Spotted garden eel	41	White-tail squirrelfish	60
Pacific flame basslet	106	Royal damsel	206	Spotted hawkfish	213	White-tip reef shark	21
Pacific mimic surgeonfish	279	Ruby cardinalfish	125	Spotted scat	173	White-tipped soldierfish	62
Pacific pinstriped butterflyfish	179	Russell's snapper	142	Spotted snake eel	40	Wide sawfish	24
Pacific sailfin tang	285	Rusty angelfish	193	Square-spot goatfish	165	Xestus sabretooth blenny	246
Pacific triangular butterflyfish	181	Saddleback hogfish	220	Squaretail grouper	96	Yellow coral goby	262
Paddlefin triggerfish	299	Saddled butterflyfish	182	Starry puffer	312	Yellow damsel	210
Painted anglerfish	50	Saddled puffer	308	Starry rabbitfish	269	Yellow devil fish	116
Painted lizardfish	45	Saddled rockcod	89	Starry triggerfish	296	Yellow hogfish	220
Painted sweetlips	150	Sailfin queen	107	Strapweed filefish	303	Yellow moon-wrasse	227
Pale surgeonfish	280	Sailfin shrimpgoby	257	Strickland's triggerfish	298	Yellow reef-basslet	110
Palette surgeonfish	278	Sailfin snapper	145	Striped catfish	43	Yellow sweeper	167
Panda clownfish	202	Sailfish	291	Striped goatfish	165	Yellow tail basslet	106
Paper fish	83	Samurai squirrelfish	59	Striped triggerfish	297	Yellow wrasse	223
Peacock flounder	292	Sandbar shark	19	Striped triplefin	243	Yellow-back fusilier	147
Peacock rockcod	90	Saowisata wrasse	222	Striped whiptail	158	Yellow-band fusilier	148
Peacock sole	293	Sargassum anglerfish	49	Surge wrasse	226	Yellow-belly damsel	209
Pearly-scale dwarf-angelfish	194	Scalloped hammerhead	22	Swallow-tail seaperch	110	Yellow-dash fusilier	148
Pencilled surgeonfish	280	Scarlet bandfish	215	Tail-blotch lizardfish	45	Yellow-dotted butterflyfish	182
Pennant bannerfish	186	Schooling bannerfish	185	Tail-spot scorpionfish	176	Yellow-edged moray	37
Peppered grubfish	241	Schooling basslets	98	Tail-spot dart goby	266	Yellow-eye puffer	312
Peron's filefish	303	Schooling harptail blenny	245	Tall-fin batfish	169, 172	Yellow-eyed combtooth blenny	247
Phantom angelfish	190	Schooling Pyramid butterflyfish	185	Tasselled wobbegong	11	Yellow-fin fairy-wrasse	231
Philippine damsel	209	Schultz's pipefish	73	Tawny shark	14	Yellow-fin flasher wrasse	232
Pickhandle barracuda	287	Scissortail dart goby	266	Teardrop butterflyfish	180	Yellow-fin soldierfish	62
Picture dragonet	252	Scribbled filefish	304	Thicklip trevally	134	Yellow-fin surgeonfish	281
Pig-face butterflyfish	180	Seba anemonefish	200	Thorny seahorse	72	Yellow-head parrotfish	237
Pin-striped wrasse	233	Sepia stingaree	31	Thorny-back cowfish	305	Yellow-lip triplefin	243
Pinjalo snapper	137	Sergeant Major	207	Thousand-spot grubfish	240	Yellow-margin triggerfish	296
Pink anemonefish	200	Shaded batfish	171	Three-colour parrotfish	236	Yellow-ribbon sweetlips	153
Pink basslet	102	Shaggy anglerfish	50	Three-line spinecheek	159	Yellow-saddle goatfish	161
Pink flasher wrasse	232	Shark ray	25	Three-spot angelfish	191	Yellow-spot goatfish	162
Pink whipray	27	Sharp-nose grubfish	241	Three-spot dascyllus	204	Yellow-spotted tilefish	128
Pink-lined goby	255	Short-snout filefish	303	Three-spot squirrelfish	60	Yellow-spotted trevally	134
Poisonous goby	260	Shortfin mako	15	Tiger cardinalfish	121	Yellow-stripe goatfish	165
Polkadot cardinalfish	126	Shortfin scorpionfish	82	Tiger shark	20	Yellow-striped emperor	156
Porcupine ray	30	Shorthead sabretooth blenny	246	Timor snapper	140	Yellow-tail parrotfish	237
Powderblue surgeonfish	278	Shortnose boxfish	307	Tomato anemonefish	199	Yellow-tail sergeant	207
Princess damsel	209	Shortnose unicornfish	276	Tomato cod	88	Yellow-tail spinecheek	160
Puntang goby	261	Silky shark	20	Tresher shark	16	Yellow-tip bristletooth	283
Purple fire goby	267	Silver batfish	173	Trumpetfish	65	Yellowfin grouper	94
Purple grouper	94	Silver puffer	314	Tube-mouth wrasse	229	Yellowfin tuna	291
Purple queen	107	Silver squirrelfish	61	Two banded soapfish	108	Yellowtail barracuda	287
Purple tilefish	129	Silver-line spinecheek	160	Two-colour parrotfish	238	Yellowtail blue-damsel	208
Purpleback dottyback	113	Silvertip shark	17	Two-eyed coralfish	183	Yellowtail scad	136
Pygmy devil ray	33	Singapore shrimpgoby	259	Two-eyed lionfish	80	Zebra batfish	171
Pyjama cardinalfish	126	Singular bannerfish	186	Two-fin flashlightfish	56	Zebra lionfish	78
Pyjamagoby	254	Six-banded angelfish	189	Two-spot basslet	101	Zebra moray	39
Queenfish	136	Six-barred wrasse	226	Two-spot bristletooth	284	Zebra shark	14
Raccoon butterflyfish	178	Skipjack tuna	288	Two-spot snapper	144		
Raggy scorpionfish	80	Slate sweetlips	150	Two-stripe featherstar clingfish	53		
Randall's basslet	100	Slender dottyback	112	Two-tone dottyback	113		
Razor wrasse	228	Slender emperor	156	Undulate moray	38		
Red bass	139	Slender grinner	46	Unicorn filefish	304		
Red cheeked basslet	104	Slender silverbelly	149	Urchin cardinalfish	126		
Red emperor	138	Slender splendour wrasse	234	V-tail tubelip wrasse	230		
Red fire goby	266	Slender sponge goby	264	Vagabond butterflyfish	181		

SCIENTIFIC INDEX

Abalistes stellatus	296	Aulostomus chinensis	65	Chaetodon oxycephalus	180	Epinephelus bleekeri	98
Ablabys macracanthus	84	Balistapus undulatus	297	Chaetodon plebeius	179	Epinephelus bontoides	96
Ablabys taenianotus	84	Balistoides conspicillum	294	Chaetodon punctatofasciatus	177	Epinephelus caeruleopunctatus	96
Abudefduf lorenzi	207	Balistoides viridescens	294	Chaetodon rafflesi	182	Epinephelus coioides	98
Abudefduf notatus	207	Belonoperca chabanaudi	110	Chaetodon reticulatus	178	Epinephelus corallicola	95
Abudefduf vaigiensis	207	Benthosema fibulatum	47	Chaetodon selene	182	Epinephelus cyanopodus	94
Acanthochromis polyacanthus	206	Bodianus anthioides	218	Chaetodon semeion	183	Epinephelus fasciatus	98
Acanthocybium solandri	291	Bodianus bilunulatus	220	Chaetodon speculum	180	Epinephelus flavocaeruleus	94
Acanthurus aurantiacavus	282	Bodianus bimaculatus	220	Chaetodon trifascialis	179	Epinephelus maculatus	95
Acanthurus bariene	281	Bodianus diana	219	Chaetodon ulietensis	181	Epinephelus ongus	97
Acanthurus dussumieri	280	Bodianus loxozonus	220	Chaetodon unimaculatus	180	Epinephelus polyphekadion	97
Acanthurus fowleri	282	Bodianus mesothorax	219	Chaetodon vagabundus	181	Epinephelus quoyanus	95
Acanthurus leucosternon	278	Bolbometopon muricatum	238	Chaetodon xanthurus	174	Epinephelus spilotoceps	97
Acanthurus lineatus	283	Bothus mancus	292	Chaetodontoplus melanosoma	190	Eucrossorhinus dasypogon	11
Acanthurus maculiceps	282	Bothus pantherinus	292	Chaetodontoplus mesoleucus	190	Eurypegasus draconis	64
Acanthurus mata	280	Brotula multibarbata	48	Chaetodontoplus septentrionalis	190	Euthynnus affinis	289
Acanthurus nigricans	280	Bryaninops loki	264	Cheilinus celebicus	234	Exallias brevis	249
Acanthurus nigricauda	281	Caesio caerulaurea	147	Cheilinus fasciatus	234	Exyrias belissimus	261
Acanthurus olivaceus	278	Caesio cuning	146	Cheilio inermis	221	Exyrias puntang	261
Acanthurus pyroferus	279	Caesio erythrogaster	146	Cheilodipterus alleni	121	Exyrias sp	261
Acanthurus triostegus	283	Caesio xanthonota	147	Cheilodipterus macrodon	121	Fistularia commersonii	66
Acanthurus tristis	279	Callionymus simplicicornis	252	Cheilodipterus quinquelineatus	121	Forcipiger flavissimus	184
Acanthurus xanthopterus	281	Callaplesiops altivelis	117	Chelmon rostratus	184	Forcipiger longirostris	184
Acentronura tentaculata	72	Canthigaster amboinensis	310	Chilomycterus reticularis	316	Galeocerdo cuvier	20
Acreichthys tomentosum	301	Canthigaster bennetti	310	Chiloscyllium griseum	12	Genicanthus lamarck	196
Aeoliscus strigatus	67	Canthigaster compressa	309	Chiloscyllium plagiosum	13	Genicanthus melanospilos	196
Aetobatus narinari	32	Canthigaster coronata	308	Chiloscyllium punctatum	12	Gerres abbreviatus	149
Alopias vulpinus	16	Canthigaster epilampra	309	Chromileptes altivelis	93	Gerres oyena	149
Aluterus monoceros	304	Canthigaster ocellicincta	309	Chromis atripectoralis	205	Gnathanodon speciosus	132
Aluterus scriptus	304	Canthigaster papua	310	Chrysiptera bleekeri	208	Gnathodentex aurolineatus	155
Amblyeleotris gymnocephala	258	Canthigaster valentini	308	Chrysiptera parasema	208	Gobiodon okinawae	262
Amblyeleotris latifasciata	258	Carangoides bajad	134	Cirrhilabrus cyanopleura	231	Gomphosus varius	226
Amblyeleotris randalli	257	Carangoides ferdau	134	Cirrhilabrus exquisitus	230	Gracila albomarginata	92
Amblyeleotris sp	258	Carangoides gymnostethus	135	Cirrhilabrus flavidorsalis	231	Grammatorcynus bilineatus	290
Amblyglyphidodon aureus	205	Carangoides oblongus	135	Cirrhilabrus solorensis	231	Grammistes sexlineatus	108
Amblygobius bynoensis	255	Carangoides orthogrammus	134	Cirrhitichthys aprinus	212	Gunnelichthys monostigma	268
Amblygobius decussatus	255	Carangoides plagiotaenia	135	Cirrhitichthys aureus	212	Gunnellichthys pleurotaenia	268
Amblygobius hectori	256	Caranx ignobilis	133	Cirrhitichthys falco	213	Gunnellichthys viridescens	268
Amblygobius nocturnus	254	Caranx melampygus	133	Cirrhitichthys oxycephalus	213	Gymnocranius microdon	157
Amblygobius sphynx	255	Caranx sexfasciatus	131, 133	Coradion chrysozonus	183	Gymnomuraena zebra	39
Amphiprion akallopisos	201	Carcharhinus albimarginatus	17	Coradion melanopus	183	Gymnorhorax melatremus	37
Amphiprion chrysopterus	200	Carcharhinus amblyrhynchos	22	Coris gaimard	225	Gymnosarda unicolor	289
Amphiprion clarkii	202	Carcharhinus galapagensis	17	Coris pictoides	225	Gymnothorax favagineus	35
Amphiprion ephippium	199	Carcharhinus leucas	20	Corythoichthys schultzi	73	Gymnothorax fimbriatus	39
Amphiprion frenatus	199	Carcharhinus limbatus	18	Crenimugil crenilabrus	217	Gymnothorax flavimarginatus	37
Amphiprion melanopus	197, 199	Carcharhinus longimanus	21	Crossosalarias macrospilus	248	Gymnothorax javanicus	37
Amphiprion ocellaris	203	Carcharhinus melanopterus	18	Cryptocentrus cyanotaenia	259	Gymnothorax meleagris	38
Amphiprion percula	203	Carcharhinus obscurus	19	Cryptocentrus polyophthalmus	259	Gymnothorax undulatus	38
Amphiprion perideraion	200	Carcharhinus plumbeus	19	Cryptocentrus singapurensis	259	Gymnothorax zonipectus	38
Amphiprion polymnus	202	Carcharhinus sorrah	18	Ctenochaetus binotatus	284	Halicampus macrorhynchus	73
Amphiprion sandaracinos	201	Carcharias falciformis	20	Ctenochaetus striatus	284	Halichoeres binotopsis	222
Amphiprion sebae	200	Carcharias taurus	15	Ctenochaetus strigosus	284	Halichoeres chrysus	223
Anampses lineatus	224	Centriscus scutatus	68	Ctenochaetus tominiensis	283	Halichoeres leucurus	222
Anampses meleagrides	224	Centropyge bicolor	192	Cyclichthys orbicularis	316	Halichoeres marginatus	222
Anomalops katoptron	56	Centropyge bispinosus	193	Cymbacephalus beauforti	86	Halichoeres trispilus	223
Anoxypristis cuspidata	24	Centropyge eibli	194	Cymolutes torquatus	228	Helcogramma gymnauchen	243
Antennarius commersonii	51	Centropyge ferrugatus	193	Cyprinocirrhites polyactis	213	Helcogramma striata	243
Antennarius hispidus	50	Centropyge flavicauda	192	Cypselurus sp	55	Hemigymnus fasciatus	221
Antennarius maculatus	51	Centropyge flavipectoralis	192	Dactyloptena orientalis	85	Hemiramphus far	55
Antennarius pictus	50	Centropyge flavissimus	191	Dactylopus dactylopus	252	Hemiscyllium strahani	13
Apogon aureus	122	Centropyge heraldi	191	Dascyllus carneus	204	Hemitaurichthys polylepis	185
Apogon bandanensis	122	Centropyge loriculus	194	Dascyllus melanurus	204	Heniochus chrysostomus	186
Apogon chrysopomus	123	Centropyge multifasciatus	195	Dascyllus trimaculatus	204	Heniochus diphreutes	185
Apogon cyanosoma	124	Centropyge tibicen	193	Dasyatis kuhlii	29	Heniochus singularius	186
Apogon dispar	122	Centropyge vroliki	194	Dendrochirus biocellatus	80	Heniochus varius	186
Apogon erythrinus	125	Cephalopholis argus	90	Dendrochirus brachypterus	78	Herklotsichthys quadrimaculatus	42
Apogon kiensis	124	Cephalopholis cyanostigma	91	Dendrochirus zebra	78	Heteroconger hassi	41
Apogon moluccensis	125	Cephalopholis formosa	91	Diademichthys lineatus	52	Heteroconger perissodon	41
Apogon multilineatus	125	Cephalopholis leoparda	90	Diagramma labiosum	150	Heterodontus japonicus	11
Apogon nigrofasciatus	124	Cephalopholis miniata	89	Diagramma pictum	150	Heteropriacanthus cruentatus	118
Apogon rhodopterus	123	Cephalopholis polleni	91	Diodon holocanthus	315	Hexanchus griseus	10
Apogon sealei	123	Cephalopholis sexmaculata	87, 89	Diodon hystrix	316	Himantura fai	27
Apolemichthys trimaculatus	191	Cephalopholis sonnerati	86, 88	Diodon liturosus	315	Himantura granulata	28
Aprion virescens	137	Cephalopholis spiloparaea	89	Diploprion bifasciatum	108	Himantura jenkinsii	27
Arothron hispidus	314	Cephalopholis urodeta	90	Discotrema echinophila	53	Himantura uarnak	28
Arothron immaculatus	312	Cephaloscyllium umbratile	16	Discotrema lineata	53	Himantura undulata	28
Arothron manilensis	312	Cepola sp	215	Doryrhamphus dactyliophorus	74	Hippocampus hixtrix	72
Arothron mappa	314	Cetoscarus bicolor	238	Doryrhamphus janssi	75	Hippocampus kuda	71
Arothron nigropunctatus	311, 313	Chaetoderma penicilligera	300	Doryrhamphus multiannulatus	74	Histrio histrio	49
Arothron reticularis	313	Chaetodon adiergastos	177	Echeneis naucrates	130	Holacanthus venustus	196
Arothron stellatus	312	Chaetodon baronessa	181	Echidna nebulosa	36	Hologymnosus doliatus	224
Aspidontus dussumieri	244	Chaetodon burgessi	174	Echidna polyzona	36	Hoplolatilus chlupatyi	129
Assessor flavissimus	116	Chaetodon citrinellus	176	Ecsenius bathi	247	Hoplolatilus fourmanoiri	128
Assessor randalli	116	Chaetodon collare	175, 178	Ecsenius melarchus	247	Hoplolatilus luteus	128
Asterropteryx ensifera	260	Chaetodon ephippium	182	Ecsenius pictus	247	Hoplolatilus marcosi	128
Atelomycterus marmoratus	16	Chaetodon lunula	178	Emiscyllium ocellatum	13	Hoplolatilus purpureus	129
Atule mate	136	Chaetodon lunulatus	179	Enchelycore pardalis	36	Hoplolatilus starcki	129
		Chaetodon melannotus	176	Enneapterygius pusillus	243	Istiblennius edentulus	248
		Chaetodon ocellicaudus	176	Epibulus insidiator	234	Istiophorus platypterus	291
		Chaetodon ornatissimus	178	Epinephelus areolatus	96	Isurus oxyrinchus	15

Name	Page	Name	Page	Name	Page	Name	Page
Katsuwonus pelamis	288	Oxymetopon cyanoctenosum	267	Pseudanthias huchtii	104	Siderea thyrsoidea	39
Kyphosus cinerascens	168	Oxymonacanthus longirostris	301	Pseudanthias hypselosoma	102	Siganus corallinus	272
Kyphosus vaigiensis	168	Oxyurichthys ophthalmolepis	260	Pseudanthias ignitus	106	Siganus guttatus	270
Labracinus cyclophthalmus	115	Paracanthurus hepatus	278	Pseudanthias lori	107	Siganus javus	270
Labrichthys unilineatus	229	Paracheilinus filamentosus	232	Pseudanthias luzonensis	102	Siganus magnificus	271
Labrobsis xanthonota	230	Paracheilinus sp 1	232	Pseudanthias pascalus	107	Siganus puellus	272
Labroides dimidiatus	230	Paracheilinus sp 2	232	Pseudanthias pleurotaenia	103	Siganus punctatus	269
Lactoria cornuta	305	Paracirrhites arcatus	214	Pseudanthias randalli	100	Siganus stellatus	269
Lactoria diaphana	306	Paracirrhites forsteri	214	Pseudanthias rubrizonatus	103	Siganus unimaculatus	271
Lactoria fornasini	305	Paraluteres prionurus	301	Pseudanthias sp.	101	Siganus vermiculatus	270
Lagocephalus sceleratus	314	Paramonacanthus curtorhynchus	303	Pseudanthias squamipinnis	98, 104	Siganus virgatus	272
Lepidichthys sp	52	Parapercis clathrata	240	Pseudanthias tuka	107	Signigobius biocellatus	253
Lethrinus microdon	156	Parapercis cylindrica	241	Pseudobalistes flavomarginatus	296	Siokunichthys nigrolineatus	75
Lethrinus ornatus	156	Parapercis millepunctata	240	Pseudobalistes fuscus	296	Siphamia versicolor	126
Lethrinus variegatus	156	Parapercis schauinslandi	241	Pseudocheilinus evanidus	233	Solenostomus cyanopterus	68
Liopropoma multilineatum	110	Parapercis xanthozona	241	Pseudochromis bitaeniatus	112	Solenostomus paradoxus	69, 70
Liza vaigiensis	217	Parapriacanthus ransonneti	167	Pseudochromis diadema	113	Sphaeramia nematoptera	126
Lutjanus argentimaculatus	140	Pardachirus melanospilos	293	Pseudochromis fuscus	114	Sphaeramia orbicularis	119, 126
Lutjanus biguttatus	144	Pardachirus pavoninus	293	Pseudochromis paccagnellae	113	Sphyraena barracuda	286
Lutjanus bohar	139	Parupeneus barberinoides	164	Pseudochromis polynemus	114	Sphyraena flavicauda	287
Lutjanus boutton	145	Parupeneus barberinus	162	Pseudochromis porphyreus	113	Sphyraena jello	287
Lutjanus carponotatus	143	Parupeneus bifasciatus	163	Pseudochromis splendens	115	Sphyraena putnamiae	286
Lutjanus decussatus	144	Parupeneus cyclostomus	161	Pseudochromis steenei	111, 114	Sphyraena qenie	287
Lutjanus ehrenbergii	142	Parupeneus heptacanthus	162	Pseudomonacanthus macrurus	303	Sphyrna lewini	22
Lutjanus fulvus	141	Parupeneus indicus	162	Pseudomonacanthus peroni	303	Sphyrna mokarran	23
Lutjanus gibbus	139	Parupeneus macronema	163	Pteragogus enneacanthus	233	Sphyrna zygaena	23
Lutjanus kasmira	143	Parupeneus multifasciatus	163	Ptereleotris evides	266	Squatina japonica	23
Lutjanus lunulatus	141	Parupeneus pleurostigma	164	Ptereleotris heteroptera	266	Stegastoma fasciatum	14
Lutjanus malabaricus	140	Pastinachus sephen	26	Ptereleotris monoptera	265	Stethojulis triliniata	223
Lutjanus monostigma	141	Pempheris adusta	166	Pterocaesio chrysozona	148	Stonogobiops nematodes	256
Lutjanus quinquelineatus	143	Pempheris mangula	167	Pterocaesio digramma	148	Sufflamen bursa	297
Lutjanus russelli	142	Pempheris vanicolensis	166	Pterocaesio randalli	148	Sufflamen chrysopterus	297
Lutjanus sebae	138	Pentapodus setosus	158	Pterocaesio tile	147	Symphorichthys spilurus	145
Lutjanus semicinctus	144	Pentapodus trivittatus	158	Pterois antennata	76, 77	Synanceia verrucosa	83
Lutjanus timorensis	140	Pervagor janthinosoma	302	Pterois miles	79	Synchiropus bartelsi	250
Lutjanus vitta	142	Pervagor melanocephalus	302	Pterois radiata	78	Synchiropus kuiteri	251
Macolor macularis	138	Pervagor nigrolineatus	302	Pterois volitans	79	Synchiropus moyeri	251
Macolor niger	138	Petroscirtes breviceps	246	Pygoplites diacanthus	195	Synchiropus ocellatus	250
Macropharyngodon ornatus	221	Petroscirtes mitratus	246	Rachycentron canadum	130	Synchiropus picturatus	252
Mahidolia mystacina	257	Petroscirtes xestus	246	Rastrelliger kanagurta	289	Synchiropus splendidus	251
Malacanthus brevirostris	127	Pholidichthys leucotaenia	242	Rhabdamia cypselura	120	Syngnathoides biaculeatus	72
Malacanthus latovittatus	127	Photoblepharon palpebratus	57	Rhina ancylostoma	25	Synodus dermatogenys	44
Manta brevirostris	33	Pinjalo pinjalo	137	Rhincodon typus	9, 10	Synodus jaculum	45
Meiacanthus atrodorsalis	245	Platax batavianus	171	Rhinecanthus cinereus	298	Synodus rubromarmoratus	44
Meiacanthus ditrema	245	Platax boersii	172	Rhinecanthus rectangulus	298	Synodus variegatus	45
Meiacanthus smithi	245	Platax orbicularis	170	Rhinecanthus verrucosus	298	Taenianotus triacanthus	83
Melichthys indicus	299	Platax pinnatus	171	Rhinomuraena quaesita	34-35	Taeniura lymna	29
Melichthys vidua	299	Platax teira	169, 172	Rhinopias aphanes	81, 82	Taeniura meyeni	30
Mobula eregoodootenkee	33	Plectorhinchus celebicus	153	Rhinoptera javanica	31	Thalassoma hardwicke	226
Monodactylus argenteus	173	Plectorhinchus chaetodonoides	151	Rhynchobatus djiddensis	25	Thalassoma janseni	227
Monotaxis grandoculis	157	Plectorhinchus flavomaculatus	153	S. orbicularis.	118	Thalassoma lutescens	227
Mulloidichthys vanicolensis	165	Plectorhinchus gibbosus	154	S. oxycephalus	80	Thalassoma purpureum	226
Muloidichthys flavolineatus	165	Plectorhinchus lessoni	152	Salarias fasciatus	249	Thalassoma quinquevittatum	227
Myersina sp.	256	Plectorhinchus lineatus	152	Salarias segmentatus	248	Thryssa baelama	43
Myrichthys colubrinus	40	Plectorhinchus obscurus	154	Sarda orientalis	288	Thunnus albacares	291
Myrichthys maculosus	40	Plectorhinchus orientalis	152	Sargocentron caudimaculatum	60	Thysanophrys otaisensis	86
Myripristis adusta	63	Plectorhinchus picus	151	Sargocentron diadema	59	Torpedo sinuspersici	26
Myripristis berndti	62	Plectorhinchus polytaenia	153	Sargocentron ittodai	59	Trachinocephalus myops	45
Myripristis melanosticta	63	Plectorhinchus sp	154	Sargocentron melanospilos	60	Trachinotus blochii	136
Myripristis murdjan	62	Plectranthias longimanus	100	Sargocentron microstoma	59	Trachyrhamphus longirostris	73
Myripristis violacea	63	Plectroglyphidodon dickii	210	Sargocentron rubrum	58	Triaenodon obesus	21
Myripristis vittata	62	Plectroglyphidodon lacrymatus	210	Sargocentron spiniferum	58	Trichonotus elegans	239
Naso annulatus	275	Plectropomus laevis	93	Sargocentron violaceum	60	Trichonotus setigerus	239
Naso brevirostris	276	Plectropomus maculatus	93	Saurida elongata	47	Trimma sp	262
Naso hexacanthus	274	Plectropomus oligacanthus	94	Saurida gracilis	46	Tylosurus crocodilus	54
Naso lituratus	277	Pleurosicya boldinghi	264	Saurida nebulosa	46	Upeneus tragula	164
Naso lopezi	275	Pleurosicya elongata	264	Scarus bowersi	236	Upeneus vittatus	165
Naso minor	276	Pleurosicya mossambica	262, 263	Scarus forsteni	236	Urogymnus asperrimus	30
Naso thynnoides	274	Plotosus lineatus	43	Scarus ghobban	235	Urolophus aurantiacus	31
Naso tuberosus	275	Pogonoperca punctata	108, 109	Scarus hypselopterus	237	Valenciennea puellaris	254
Naso unicornis	276	Pomacanthus annularis	188	Scarus niger	237	Valenciennea randalli	254
Naso vlamingii	277	Pomacanthus imperator	188	Scarus rubroviolaceus	235	Valenciennea wardii	253
Nebrius ferrugineus	14	Pomacanthus navarchus	187, 189	Scarus sp	238	Variola albimarginata	92
Nemateleotris decora	267	Pomacanthus semicirculatus	188	Scarus spinus	237	Variola louti	92
Nemateleotris magnifica	266	Pomacanthus sexstriatus	189	Scarus tricolor	236	Wetmorella albofasciata	233
Neoglyphidodon melas	206	Pomacanthus xanthometopon	189	Scatophagus argus	173	Xanthichthys auromarginatus	299
Neoglyphidodon nigroris	205	Pomacentrus auriventris	209	Scolopsis affinis	160	Xiphasia setifer	244
Neoniphon argenteus	61	Pomacentrus moluccensis	210	Scolopsis aurata	160	Xyrichtys aneitensis	229
Neoniphon opercularis	61	Pomacentrus philippinus	209	Scolopsis ciliata	160	Xyrichtys pavo	228
Neoniphon sammara	61	Pomacentrus vaiuli	209	Scolopsis temporalis	159	Xyrichtys pentadactylus	229
Neopomacentrus anabatoides	208	Premnas biaculeatus	198	Scolopsis trilineata	159	Yongeichthys nebulosus	260
Novaculichthys taeniourus	228	Priacanthus blochii	118	Scolopsis xenochroa	159	Zanclus cornutus	273
Opistognathus sp	216	Prionace glauca	21	Scomberoides lysan	136	Zebrasoma scopas	285
Orectolobus ornatus	12	Pristis pectinata	24	Scomberomores semifasciatus	290	Zebrasoma veliferum	285
Ostracion cubicus	307	Pseudaluterus nasicornis	304	Scomberomorus commerson	290		
Ostracion meleagris	306	Pseudanthias bicolor	100	Scorpaenodes parvipinnis	82		
Ostracion nasus	307	Pseudanthias bimaculatus	101	Scorpaenodes varipinnis	82		
Ostracion rhinorhynchus	307	Pseudanthias dispar	106	Scorpaenopsis venosa	80		
Ostracion solorensis	306	Pseudanthias evansi	106	Seriola rivoliana	132		
Oxycirrhites typus	214	Pseudanthias fasciatus	102	Serranocirrhitus latus	110		

FISH-FEATURES

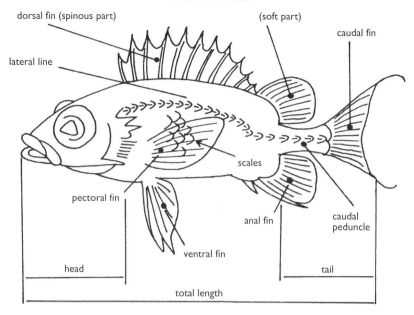

GLOSSARY

axil: inner base of pectoral fin; **barbel:** fleshy sensory appendage on the head, usually on the chin or near mouth; **benthic:** occurs on the seafloor; **cryptic:** pertaining to concealment by colour and behaviour; **demersal:** living on the seafloor; **detritus:** organic matter, primarily plant material and faeces; **diurnal:** active during the day; **endemic:** found only in a given limited region; **intertidal:** portion of the shoreline that is between the lowest and the highest tides; **nocturnal:** active during the night; **pelagic:** in the open water, well above the bootom; **thermocline:** zone of rapidly dropping temperature with increased depth.

PHOTO CREDIT

Cover left to right:
Two whalesharks	Foto: Mark Strickland
Two-spot basslet	Foto: Rudie Kuiter
Flame dwarf-angelfish	Foto: Helmut Debelius
Jenkins whipray	Foto: Mark Strickland
Purple fire goby	Foto: Helmut Debelius
Magenta dottyback	Foto: Helmut Debelius

Backcover:
Zebra batfish   Foto: Rudie Kuiter

Single Photo credit:
(o: upper, m: center, u: lower, k: small)

FRED BAVENDAM:	30o, 56m, 82ok.
JÜRGEN BREI:	30u, 30uk.
CHRISTOPH GERIGK:	21o.
HOWARD HALL:	11u, 15o, 15u, 19o.
ARNE HODALIC:	119.
AVI KLAPFER:	22u, 22uk, 33uk, 251m, 273u, 291u.
SCOTT MICHAEL:	12u, 13o, 13mk, 13u, 13uk.
DOUG PERRINE:	18m, 20m, 23o, 31u.
JAN POST:	18o, 19u, 113m, 114o, 115u.
ED ROBINSON:	22o, 26u, 29o, 33o, 33ok, 33u, 35o, 85u, 171u, 274m, 275u, 280u.
MARK STRICKLAND:	9, 10u, 14o, 14ok, 14m, 17o-u, 25o-u, 27o-u, 28o, 28ok, 32o, 37o, 37u, 52u, 70o, 73m, 74u, 82u, 89mk, 91m, 99, 132m, 137u, 175, 216m, 243uk, 291o, 298o.
PHIL WOODHEAD:	26o, 28m, 79u.
NORBERT WU:	11ko, 24m, 33uk.

All other photos were taken by the authors.

TROPIC OF CANCER

Hainan
Senck.

THAILAND

Bangkok

VIETNAM

Saigon

Andaman Is.
Port Blair

ANDAMAN SEA

SOUTH CHINA SEA

Similan Is.
Phuket

Sabang Is.

Kota Kinobalu

Medan

MALAYSIA

Kuala Lumpur

Singapore

Borneo

Sumatra

Kalimantan

EQUATOR

Padang

Balikpapan

INDO

JAVA SEA

Jakarta

Java

Bali

INDIAN OCEAN

Christmas Is.

Cocos Keeling Is.